12th International Conference on Advances in Mass Data Analysis of Images & Signals (MDA 2017)

With Applications in Medicine, r/g/b Biotechnology, Food Industries and Dietetics, Biometry and Security, Agriculture, Drug Discover, and System Biology Medicine, Biotechnology, Chemistry and Food Industry

New York, New York, USA
8 - 11 July 2017

Editor:

Petra Perner

ISBN: 978-1-5108-8533-2

Printed from e-media with permission by:

Curran Associates, Inc.
57 Morehouse Lane
Red Hook, NY 12571

Some format issues inherent in the e-media version may also appear in this print version.

ibai-publishing
Arno-Nitzsche-Str. 45
04277 Leipzig, Germany

info@ibai-publishing.org

Additional copies of this publication are available from:

Curran Associates, Inc.
57 Morehouse Lane
Red Hook, NY 12571 USA
Phone: 845-758-0400
Fax: 845-758-2633
Email: curran@proceedings.com
Web: www.proceedings.com

Petra Perner (Ed.)

Advances in Mass Data Analysis of Images and Signals

with Applications in Medicine, r/g/b Biotechnology, Food Industries and Dietetics, Biometry and Security, Agriculture, Drug Discover, and System Biology

12th International Conference, MDA 2017
New York, USA, July 2017

Proceedings

ibai Publishing

www.ibai-publishing.org

The German National Library listed this publication in the German National Bibliography.
Detailed bibliographical data can be downloaded from http://dnb.ddb.de.

ibai-publishing
Prof. Dr. Petra Perner
Arno-Nitzsche-Str. 45
04277 Leipzig, Germany
E-mail: info@ibai-publishing.org
http://www.ibai-publishing.org

Printed in Germany, 2017

Table of Contents

Preface

The 12th International Conference on Mass Data Analysis of Signals and Images with Applications in Medicine, r/g/b Biotechnology, Food Industries and Dietetics, Biometry and Security, and Agriculture MDA 2017 held in July in Newark USA showed once more that the event is a must for all specialists from research and industry alike who like to stay informed about hot new topics in mass data analysis of signals and images.

We accepted nine papers on very important topics. The acceptance rate was 33%. The topics range from Agriculture Signal Analysis, EEG Signal Analysis, Imageing in Fish Inspection, Telemedicine, OMR Scanning, Electronic Device Analysis, Handwriting Recognition, Case-Based Object Recognition, and Hazy Image Analysis.

A tutorial on Standardization of Immunofluorescence and a tutorial on Intelligent Image Interpretation has been held in connection with the conference.
Selected revised papers of this conference as well as contributed papers will be published in the October 2017 issue of the International Journal Transaction on Mass Data Analysis of Signals and Images (www.ibai-publishing.org) that has ***an impact factor of 0.533.***

We are happy to see that new automatic systems for the analysis of signals and images have been presented at MDA 2017 and that a lot of researchers followed our mission to bring new signal and image analysis methods into real applications that introduce a new quality level to various real life applications.

The 13th conference will be held in 2018 in New York (www.mda-signals.de) under the auspices of the World Congress Frontiers on Intelligent Data Analysis DSA 2018 (www.worldcongressdsa.com).
We would like to invite you to contribute to this conference. Please come and join us.

July, 2017

Prof. Dr. Petra Perner
Chair of the Conference MDA

International Conference on Mass Data Analysis with Applications in Medicine, r/g/b Biotechnology, Food Industries and Dietetics, Biometry and Security, and Agriculture MDA 2017

Chair

Petra Perner	IBaI Leipzig, Germany

Committee

Walter Arnold	Fraunhofer-Institute of non-destructive Testing, Germany
Valentin Brimkov	Buffalo State College, USA
Hans du Buf	University of Algarve, Portugal
Calin Ciufudean	"Stefan cel Mare" University, Romania
Kamil Dimililer	Near East University, Cyprus
George C. Giakos	University of Akron, USA
Giulio Iannello	University Campus Bio-Medico of Rome, Italy
Rainer Schmidt	University of Rostock, Germany
Paolo Soda	University Campus Bio-Medico of Rome, Italy
Thang Viet	Pham Intelligent Systems Lab Amsterdam (ISLA), The Netherlands
Patrick Wang	Northeastern University, USA

Machine Learning Algorithms for Predicting Land Suitability in Crop Production: A Review

Kennedy Mutange Senagi[1,2], Nicolas Jouandeau[2], Peter Kamoni[3]

[1]Dedan Kimathi University of Technology, Kenya

[2]Université Paris8, France

[3]Kenya Agricultural and Livestock Research Organization, Kenya

Abstract. Food security is an important factor to consider for a healthy nation. Agriculture is a an important sector for Kenya's economic growth and for achieving Kenya's Vision 2030. However, due to the change in farming practices from one season to the other, increased demand for land, intensive land use and failure to apply recommended farming practice results in soil degradation over time. Land evaluation is the process of assessing land performance for specified purposes *e.g.* crop production. Currently, in the Department of Soil Survey at Kenya Agricultural and Livestock Research Organization, and in many other soil survey institutions, land evaluation process is done manually and is time consuming, stressful and prone to human errors. In this paper, we review Machine Learning (ML) algorithms applied in land suitability for crop production and performance comparison of ML algorithms. We found out that parallel random forest (PRF) performed better that other supervised machine learning classification algorithms. We set up an experiment prototype, PRF scored an average RMSE of 1.03, ACC score of 0.92 and average execution time of 1532.32 milliseconds. PRF can be applied in predicting land suitability for crop production. This can reduce human errors, reduce time, and improve accuracy in land evaluation process for crop production. machine learning, parallel random forest, land evaluation, soil analysis.

1. Introduction

1.1. State of Food Security in Kenya

Kenya heavily relies on agriculture for economic growth and food security [1]. Agriculture contributes about a quarter of Kenya's Gross Domestic Product (GDP). The baseline of agriculture is expected to grow at approximately 3.7% per year during

2010-2020. The Kenyan government total public expenditure on agriculture sector has been increasing in recent years. This is to accelerate growth in the agricultural sector, increase food production, lower food prices and reduce global economic crisis [2]. In Kenya Vision 2030, agriculture is one of the sectors identified to contribute to the realization of the 10 per cent economic growth rate per annum envisaged under the economic pillar. In Vision 2030 document, Kenya desires to promote an innovative, commercially oriented and modern agriculture sector [3].

1.2. Land Evaluation for Crops Suitability

Due to changes in agricultural practices and increased demand of land, land evaluation has come to be a very important procedure before initiating land use in a particular land *e.g.,* engaging in farming activities on a piece of land or urban planning [4]. Land evaluation is the assessment of performance of Land for a defined use. The principal objective of land evaluation is to select the optimum land use for each defined land unit, taking into account both physical and socio-economic considerations as well as the conservation of environmental resources for future use [5]. It involves the analysis and interpretation of basic surveys of: soil, vegetation, climate, socio-economic (if available) and other aspects of land such as crop land use requirements. The beneficiaries of land evaluation process are the farmers who gain knowledge of what crops to grow in a particular piece of land, improve on their crop yields as well as learning new skills for sustainable land management [6]. Land evaluation, involves matching land areas called Land Mapping Units (LMU) with land uses, called Land Utilization Types (LUT), to determine the relative suitability of each land area for each land use. Land Utilization Types are specified by a set of Land-Use Requirements (LUR), which are the conditions of land necessary for the successful and sustained practice of a given LUT [6].

In practice, representative soil profiles are dug in each LMU and sampled horizon wise. The soil samples are analyzed in the laboratory for physical and chemical composition. Data from such soil profiles together with climatic, crop requirement and socio-economic data (if available) are then used to perform the land evaluation. The process involves selecting the land qualities/characteristics to be used (*e.g.*, temperature regime, soil water holding capacity, oxygen availability, rooting depth and salinity hazard) and rating them. The LMU are rated based on the selected rating scheme. The same rating scheme is used to rate the crop requirements. A matching scheme of the rated LMU qualities/characteristics with rated crop land use requirements is applied to determine the land suitability of that LMU for the specific crop. Suitability of a LMU can be categorized as either S1 (highly suitable i.e. where 75% to 100% yield is attained), S2 (moderately suitable *i.e.,* where 50% to 75% yield is attained), S3 (marginally suitable *i.e.* where 25% to 50% yield is attained) or NS (unsuitable *i.e.*, where below 25% yield attained) [7].

Land suitability is the ability of a portion of land to allow the production of crops in a sustainable manner. The analysis identifies the main limiting factors for a particular crop production. It also enables decision makers to develop a crop management system for increasing the land productivity or choose an appropriate fertilizer to increase productivity of the land [6].

The suitability of each LMU for each LUT is done as follows [6]:

1. Determining the actual Land Characteristic (LC) values for the LMU. These can be measured, observed in the field and estimated through laboratory measurements or remote sensing
2. Combining the Land Characteristics (LC) values into Land Quality (LQ) values (*i.e.*, inferring the LQ from the set of LC's)
3. Matching the LQ values with Land Use Requirements *e.g.*, crop requirements
4. Inferring the suitability from the set of LQ's and LURs)

1.3. Computing Techniques in Land Evaluation

In recent years, computing concepts have been successfully applied in land evaluation for crop suitability. Some of the computing techniques applied are: Geographical Information Systems (GIS) and machine learning. GIS are computer based tools that captures, stores, analyses, manages and presents all types of spatial or geographical data. GIS techniques have successfully been used to evaluate land suitability [24]. Machine learning is applied in various land evaluation processes which include: predicting soil properties [8], predicting soil fertilizer requirements [9] and land use suitability [10]. In essence, machine learning is concerned with the question of how to construct computer programs that automatically improve with experience. Besides agriculture, state of the art machine learning applications have successfully been developed, for examples, data-mining programs that learn to detect fraudulent credit card transactions, information-filtering systems that learn a users' reading preferences and autonomous vehicles that learn to drive on public highways [11].

2. Machine Learning

Machine learning is a discipline in computer science that emerged from the field of artificial intelligence. Machine learning is concerned with the construction of computer programs that automatically improve with experience. A computer program is said to learn from experience E with respect to some class of tasks T and performance measure P, if its performance at tasks in T, as measured by P, improves with experience E. It is too costly and impractical to design intelligent systems by gathering experts' knowledge and then hard-wiring it into a machine. To build intelligent machines, researchers have comprehended that these machines should learn from and adapt to the environment they learn in. Building intelligent machines involve learning from data [12]. Common types of machine learning problems are categorized as: supervised machine learning, unsupervised machine learning and reinforcement learning.

2.1. Supervised Machine Learning

In Supervised ML algorithms, a general hypothesis is built from external supplied data. From the hypothesis we can then make predictions of future instances. Actually, in Supervised ML, a concise model of distribution of class labels is built from predictor features. This classifier is then used to assign class labels (unknown) to predictor features (known) [33]. Supervised learning problems can be categorized as: classification and regression. They are discussed below:

1. Classification Problem: The goal of classification is to learn a mapping of inputs x to outputs y, where $y \in \{1, \ldots, C\}$, with C being the number of classes. If $C = 2$, then it is a binary classification problem, if $C > 2$, then it is a multi-class classification problem. In classification problems, we assume $y = f(x)$ for some unknown function f, the goal of learning is to estimate the function f given a labeled training set, and to make predictions using $\hat{y} = \hat{f}(x)$. The main goal is actually to make predictions on new inputs that we have not seen before *i.e.* generalization. Some examples of classification supervised machine learning algorithms are: decision trees, ensembles (bagging [21] random forest [20] boosting [26]), k-NN and neural networks [13].

2. Regression Problem: Regression problems are just like classification, however, the output consists of one or more continuous variables. For example, predicting tomorrow's stock market price given the current market conditions and other relevant information. Examples of regression machine learning algorithms include: naive bayes and logistic regression [14].

2.2. Unsupervised Machine Learning

In unsupervised learning problem, we are given output/labeled data, without any inputs the goal is to learn relationships and structure from such data, this is sometimes called knowledge discovery. Unlike supervised learning, the training data consists of a set of input without any corresponding output values. Unsupervised learning problems can be formalized as one of density estimation, i.e. we want to build models of the form $p(x_i|\theta)$, unlike supervised learning which is $p(y_i|x_i, \theta)$. Supervised learning is a conditional density estimation, while unsupervised learning is an unconditional density estimation. By contrast, in unsupervised learning x_i is a vector of features and we need to create multivariate probability models, whereas, in supervised learning, y_i is single variable that we are trying to predict. Therefore, supervised learning problems can use univariate probability models. Examples of unsupervised machine learning algorithms are: k-means and Neural Networks [13].

2.3. Reinforcement Machine Learning

Reinforcement learning [15] is concerned with finding suitable actions to take in a given situation in order to maximize a reward. In contrast, with supervised learning which are given labeled data, reinforcement learning algorithms discover optimal outputs by a process of trial and error. Generally, there is a sequence of states and actions in which the learning algorithm is interacting with its environment. In many cases, the current action, not only affects the immediate reward, but also has an impact on the reward at all subsequent time steps [12].

3. Machine Learning Algorithms Applied to Land Evaluation for Suitability Assessment

In Anitha and Acharjya (2017) studies, their research objective was to predict crop to be grown in Vellore District. They came up with a hybridizing of rough set on fuzzy approximation space and ANN. The rough set on fuzzy approximation space was to get almost equivalence classes where attribute values are not qualitative. The classified information was the taken to ANN algorithm for training, prediction and testing. They used a dataset collected from Krishi Vigyan Kendra of Vellore District between 2007 to 2014. The dataset contained 2193 objects with 15 attributes. They used 26 soil attributes: soil pH, moisture, organic matter, availability of nitrogen, availability of phosphorus, availability of potassium, water pH, calcium, nitrate, magnesium, selenium rainfall, copper, zinc, manganese and iron. The dataset was divided into 55% training and 45% testing data and validated by N-folds cross-validation. The experiments were developed in R language. The average Mean Square Error (MSE) was 0.2436 and Accuracy of 93.2% [28].

Fereydoon *et al.* (2014) developed a Support Vector Machine (SVM) - Two Class Model for land suitability analysis for wheat production in Kouhin region in Iran. They used twenty two soil representative soil profiles information, each soil profile having ten features: climatic (precipitation, temperature), topographic (relief and slope) and soil-related (texture, $CaCO_3$, Organic Carbon, coarse fragment, pH, gypsum). They implemented a Two Class SVM model on a non-linear class boundaries; non-linear mapping of input vector into a high dimensional feature space. They used MATLAB 8.2 software to design and test the SVM model. They randomized the dataset and split it into training (80%) and test (20%). In performance evaluation, they got an RMSE of 3.72 and R^2 of 0.84 [4].

Land suitability classification using a large number of parameters is time consuming and costly. With this research problem, Hamzeh *et al.* (2016) presented a combination of feature selection (best search, random search and genetic search methods) and fuzzy-analytical hierarchical process(AHP) methods to improve selection of important features from a large number of parameters. On feature selection, random search performed slightly better than genetic search methods and best search. The dataset was retrieved from land classification report published by Khuzestan Soil and Water Research Institute. They found that soil texture, wetness, salinity and alkalinity were the most effective parameters for determining land suitability classification for the cultivation of barely in the Shavur Plain, southwest Iran. The report showed that soil salinity and alkalinity, soil wetness, $CaCO_3$, gypsum, pH, soil texture, soil depth and topography were the most important soil properties to consider for cultivating barley in the study area [29].

Mokarram *et al.* (2015) used AI and ML to automate the land suitability classification for growing wheat. They used a dataset with data collected from Shavur plain, northern of Khuzestan province, southwest of Iran. The dataset had the following attributes: topography (Primary slope, Secondly slope and Micro relief), salinity and alkalinity (EC and ESP), wetness (Groundwater depth and Coroma depth), soil texture, $CaCO_3$, soil depth, gypsum and pH (H_2O). Land suitability classes were classified as highly suitable (75%-100%), moderately suitable (50%-75%), marginally suitable (25%-50%) and not suitable (25%-0%). Mokarram et al. (2015 implemented Bagging, AdaBoost and RotForest algorithms and evaluated the performance of their experiments using 632+ bootstrap and 10-fold cross validation. Their results classified 26% of the land being moderately suitable, 25% being marginally suitable and 49% being not suitable. Moreover, they found that RotBoost algorithm had a better accuracy than Single Tree, Rotation Forest, AdaBoost, Bagging algorithms in predicting land suitability class. RotBoost recorded 99% and 85% accuracy score for bootstrap and cross validation respectively [30].

Dahikar and Rode (2014) demonstrated the use of ANN in predicting crop suitability from soil attributes. The attributes include: type of soil, pH, nitrogen, phosphate, potassium, organic carbon, sulphur, manganese, copper, calcium, magnesium, iron, depth, temperature, rainfall, humidity. They set up the experiments in MatLab [31]. However, performance results were not given in the papers. The researchers acknowledge the potential of using ANN in predicting crop suitability from soil data collected from rural district.

Elsheikh *et al.* (2013) presented ALSE, which was an intelligent system for assessing land suitability for different crops (*e.g.,* mango, banana, papaya, citrus, and guava) in tropical and subtropical regions based on geo-environmental factors. ALSE supported GIS capabilities on the digital map of an area with FAO-SYS framework model. It also had some necessary modifications to suit the local environmental conditions for land evaluation, and the support of expert knowledge. ALSE had the capability to identify crop-specific conditions and systematically computes the spatial and temporal data with maximum potential. This would help land planners to make complex decisions within a short period taking into account sustainability of a crop. Test dataset was collected from agricultural land in Terengganu, West Malaysia. Some of the attributes they used were: rainfall, soil attributes (nutrient availability, rooting conditions, nutrient retention, soil workability and oxygen soil drainage class) and topology [32].

From the above literature, we see diverse applications of statistical and ML techniques in land suitability for better crop production: ANN, AdaBoost, SVM, fuzzy logic and PCA. Moreover, algorithms perform differently on different datasets. Experiments set up comparing different ML algorithms on standardized dataset can be better analysis.

4. Machine Learning Algorithms Comparison and Analysis

Besides ML being applied in agriculture, we found that researches who analyzed ML algorithms and measured their performance on standardized experiments. These analysis can guide to select an algorithm that is robust and has better performance. For example, Rich *et al.* (2008) did an empirical evaluation of supervised learning of high dimension data and evaluated performance on three metrics, namely: accuracy (ACC), Area Under the ROC Curve (AUC), and squared error (RMS). They also studied the effect of increasing dimensionality, ranging from 761 to 685569, the performances of the algorithms: SVM (LaSVM kernel and RBF kernels using stochastic gradient descent), ANN, Logistics Regression (LR) and Naïve Bayes (NB). Across all dimensions, RF was the best in performance, followed by ANN, then boosted trees, and lastly SVMs [18]. These results were similar to Ogutu *et al.* (2011) that showed RF outperforming SVM [23]. However, if results of each metric are examined, in terms of accuracy, boosted decision trees are the best performing models followed by RF. It's worth noting that boosted decision trees performed better than RF in relatively low dimensionality, but as dimensionality increases RF tends to perform much better than boosted decision trees. In terms of RMS performance metric, RF is marginally better than boosted trees. While in AUC, RF is a clear winner, followed by k-NN. It's important to note that, although ANN performs second best overall, it's neither first or second in either performance metrics. This is because, ANN consistently produce very good results, though perhaps not outstanding on all the performance metrics.

Kim *et al.* (2012) experiments on image classification showed that SVM performance was much better than k-NN [16]. Similarly, Sanghamitra *et al.* (2011) compared the performance of SVM and k-NN on Oriya Character Recognition and found SVM with an accuracy rate of 98.9% while k-NN had 96.47%. Thus showing that SVM had a better accuracy rate, though k-NN classifier consumed lesser storage space and had less computation than SVM [17]. SVMs performed generally better and had more accurate results than Artificial Neural Networks (ANNs) though difficult to understand the learned function. Moreover, Colas and Brazdil (2006) experiment on text classification showed that k Nearest Neighbour (k-NN) could scale up very well and even achieve better results than SVM [19].

Similarly, Manuel *et al.* (2014) evaluated 179 classifiers got from 17 families namely: discriminant analysis, multiple adaptive regression splines, Bayesian, rule-based classifier, boosting, neural networks, bagging [21] stacking, random forests and other ensembles, SVM, decision trees, generalized linear models, logistic and multinomial regression, nearest neighbours, partial least squares and principal component regression and other methods. They were implemented in Weka, Matlab, R (with and without the caret package) C and other relevant classifiers. Parallel random forest, a version of RF, turned out to be the best classifier when implemented in R and accessed without caret. It achieved 94.1% accuracy. The second best was SVM with Gaussian and polynomial kernels when implemented in C using LibSVM which achieved 92.3% accuracy. Moreover, different learning algorithms portrayed different performance levels on change in dimensionality. Increase in dimensionality, affect the relative performance of the learning algorithm. On high dimensional data, RF performs very well followed by boosted trees and logistic regression whereas boosted decision trees perform exceptionally well when dimensionality is low [22].

Moreover, Hastie *et al.* 2009 did a comparison of characteristics of ANN, SVM, trees and k-NN algorithms. Tress showed a good performance in: natural handling data mixed type, handling data that has mixed values, robustness to outliers, insensitivity to monotone transformation of inputs, computational scalability and ability to deal with irrelevant inputs. k-NN was the second best followed by SVN and ANN which performed fairly better in: ability to extract linear, combinations of features, interpretability and predictive power. This shows that Machine learning algorithms perform differently in differently in different experiment environments. Some of the factors that affect performance of machine include: number of records in a data, type of data (*e.g.*, images, text, voice) and complexity of data [22]. Manuel *et al.* (2014) measured the percentage of the maximum accuracy of several algorithms including parallel RF, RF, rotational forest, SVM, k-NN and ANN. They found out that parallel random forest scored a better maximum accuracy than the others.

This research paper has discussed some of the machine learning techniques applied to land evaluation and crop suitability assessment. We have also discussed ML algorithms and applications. We see that parallel random forest performs comparatively well compared to SVM, ANN, k-NN and other machine learning classification algorithms.

5. Experiment Prototyping

5.1. Methodology

The standardized dataset [25] was tailored to predict the age of abalone tree from the attributes: sex, length, diameter, height, whole weight, shucked weight, viscera weight, shell weight and rings. Some of the reasons that led us to select the dataset to prototype parallel random forest were: the number of attributes/features, size of dataset and the dataset is intended for classification. These are characteristics that we expect to find in soil analysis dataset [4].

In this study we implemented the experiment prototype in Python 3.5 in Ubuntu 16.04 LTS 64 bit operating system running on Intel® Core™ i3-2365M CPU @ 1.40GHz × 4 computer systems. We randomized the data and used two thirds of the 4177 records for training the model and one third for as the test data. We run each experiment 10 times and recorded the duration of each run, the ACC score, standard deviation (for y and $\hat{y}$) and RMSE. We used RMSE defined in equation (1) and ACC score defined in equation (2), where $\hat{y}$ is the predicted value, y is the actual value, n is the number or records in the test or predicted dataset and m is the number of runs. RMSE is used to measure the accuracy and validity of the predicted values. RMSE has been used to measure predictability accuracy. Standard deviation(std) of the y training set, $\hat{y}$ of the predicted and all the y labels results was captured [4]. Moreover, we used Speedup to evaluate the performance of parallelizing RF. Speedup measure how much performance gain is achieved by parallelizing RF, Speedup is defined in

$$Average\ RMSE_j = \frac{1}{m}\sum_{m=1}^{m=10} \sqrt{\frac{1}{n}\sum_{i=1}^{n} \left(y_i^{(j)} - \hat{y}^{(j)}\right)^2} \tag{1}$$

$$Accuracy\ score(y, \hat{y}) = \frac{1}{n_{samples}} \sum_{i=0}^{n_{samples}-1} 1 - (\hat{y}_i = y_i) \tag{2}$$

5.2. Results

Results were tabulated in Table **1**. We observed RF and PRF had the same RMSE, ACC and standard deviations. RF recorded a significantly high execution time compared to PRF.

Table 1. Average results for Parallel Random Forest and Random forest algorithms.

Algorithm	RMSE	ACC	Std of $\hat{y}$ (pm)	Std of y (pm)
RF	1.03	0.92	2.45	3.3
PRF	1.03	0.92	2.45	3.3

Table 2. Parallization of Random Forest algorithm.

Core(s)	Time (ms)
1	4316.0
2	3660.0
4	3159.0
8	2769.0
16	2912.0
24	3041.0
32	3135.0
40	3247.0
48	3245.0

5.3. Discussion

RMSE measures the accuracy of the predicted values and the actual value. The best RMSE value is zero; the higher the value, the lower the accuracy. The prototype experiment scored an average value of 1.03 meaning the RMSE was averagely good. ACC computes the subset accuracy. The best ACC score is 1 and the worse is 0. The results show that both PRF and RF had an ACC score of 0.92 meaning the score was significantly good. Both RF and PRF scored a standard deviation (Std) of 2.45. The best std is zero. The higher the std, the higher the deviation from the mean. Meaning the data ranges of the y labels were deviating significantly from the mean. The results of RMSE, ACC score and std could be due to the nature of the dataset(high/low biased data), complexity of the records or number of features(high/low) [22].

Moreover, RF recorded a higher execution time than PRF. Meaning PRF was faster than RF. This could be because PRF utilized multiple CPU cores compared to RF that used a single core. Logically, we expected to get more work done by N processors to be N times faster, which was not the case as seen in Table **2**. This could be due to a certain amount of overhead incurred in keeping the parallelized system working. Some of the overheads include: time spent in interprocess interaction and idling time as a result of load imbalance and/or synchronization [27].

6. Conclusion

Kenya is keen in food security in order to feed her population. Land evaluation for crop suitability assessment is an important procedure for identifying crops that grow optimally in a specific soil sample. In Kenya Soil Survey Department, Kenya Agricultural and Livestock Research Organization and many other institutions, land evaluation is done manually, time consuming, stressful and prone to human errors. Making the process tedious and prone to errors. Computer science techniques have been proposed to leverage the challenges including geographical information systems and machine learning. We reviewed literature of applications of ML algorithms in land suitability evaluation for crop production. From the review, we found that PRF performed comparatively well compared to other algorithms like k-NN, ANN and SVM. PRF had a good handling of data of mixed type and values, was robust to handle outliers in an input space, had a good predicative accuracy and could be scaled to large datasets. We set up a simple experiment prototype to implement PRF. PRF had a significant performance in terms of RMSE, Accuracy and Time of Training. Among other machine learning preparation procedures, we need to: have sufficient number of records for training the model, take note of data complexity have enough number of features in the dataset. In essence, this prototype can be scaled up to land suitability dataset, as we intend to utilize machine learning algorithms in soil analysis to optimize crop production.

References

1. Ngeno V, Mengist C, Langat B. (2011). Technical efficiency of maize farmers in Uasin Gishu district, Kenya. 10th African Crop Science Conference Proceedings, Maputo, Mozambique, pp. 41-47 ref.25.
2. Athur M, Karl P, Samuel B. (2012). Regional Strategic Analysis and Knowledge Support System (ReSAKSS). ReSAKSS Africa, Washington.
3. Government of Kenya. (2007). A Globally competitive and prosperous Kenya. www.opendata.go.ke.
4. Fereydoon S, Ali K, Azin R, Ghavamuddin Z, Hossein J, Munawar I. (2014). Support vector machines based-modeling of land suitability analysis for rainfed agriculture. Journal of Geosciences and Geomatics, Vol. 2, Issue No. 4, pp. 165-171.
5. FAO. (1985). Guidelines for land evaluation for rainfed agriculture. FAO, Rome.
6. FAO. (1976). FAO And Agriculture Organization Of The United Nations. FAO, Rome.
7. Abbas T, Fereydoon S. (2015). Qualitative land suitability evaluation for maize (Zea mais L.) in Abyek, Iran using FAO method. Global Journal of Research and Review (GJRR), Vol. 2, Issue No. 1, pp. 37-44.
8. Jay G, Anurag I, Jayesh G, Shailesh G, Vahida A. (2012). Soil data analysis using classification techniques and soil attribute prediction. International Journal of Computer Science Issues, Volume 9, Issue No. 3.
9. Gholap J. (2012). Performance tuning of J48 algorithm for prediction of soil fertility. Asian Journal of Computer Science and Information Technology, Vol 2, Issue No. 8.
10. Shattri BM, Saied P, Ahmad RBM, Saied P. (2012). Optimization of land use suitability for agriculture using integrated geospatial model and genetic algorithms. ISPRS Ann Photogramm Remote Sens Spat Inf Sci I-2, pp. 234-299.
11. Welling M. (2011). A First Encounter with Machine Learning. Irvine, CA.: University of California.

12. Jordan M, Kleinberg J, Schölkopf B. (2007). Information science and statistics. Springer Science+Business Media, LLC.
13. Murphy KP. (2012). Machine learning a probabilistic perspective. London, England: The MIT Press.
14. James G, Witten D, Hastie T, Tibshirani R. (2013). An introduction to statistical learning. Springer (Vol. 6). New York: Springer.
15. Sutton RS, Barto AG. (1998). Reinforcement learning: An introduction. Cambridge: MIT press, Vol. 1, Issue No. 1.
16. Kim J, Byung-Soo K, Silvio S. (2012). Comparing image classification methods: K-nearest-neighbor and support-vector-machines. American-Math' 12/CEA' 12 Proceedings of the 6th WSEAS International Conference on Computer Engineering and Applications, and Proceedings of the 2012 American conference on Applied Mathematics, ISBN: 978-1-61804-064-0, pp. 133-138.
17. Sanghamitra M, Himadri NDB. (2011). Performance comparison of SVM and k-NN for oriya character recognition. International Journal of Advanced Computer Science and Applications, Special Issue on Image Processing and Analysis, pp. 112-116.
18. Rich C, Alexandru N-M. (2006). An empirical comparison of supervised learning algorithms. In Proceedings of the 23rd international conference on Machine learning, ACM, pp. 161-168.
19. Colas F, Brazdil P. (2006). Comparison of SVM and some older classification algorithms in text classification tasks. In IFIP International Conference on Artificial Intelligence in Theory and Practice, pp. 169-178, Springer US.
20. Breiman L. (2001). Random forests. Machine learning, Kluwer Academic Publisher, 45(1), pp. 5-32.
21. Breiman L. (1996). Bagging predictors. Machine Learning, Kluwer Academic Publishers, Boston, Vol 24, Issue No. 2, pp. 123-140.
22. Manuel FD, Eva C, Senen B. (2014). Do we need hundreds of classifiers to solve real world classification problems?. Journal of Machine Learning Research, pp. 3133-3181.
23. Ogutu JO, Piepho HP, Schulz-Streeck T. (2011). A comparison of random forests, boosting and support vector machines for genomic selection. In BMC proceedings, Vol. 5 (Suppl 3), pp. 1-5, BioMed Central.
24. Abraham M, Daniel H, Abeba N, Tigabu D, Temesgen G, Hagos G. (March, 2015). GIS based land suitability evaluation for main irrigated vegetables in Semaz Dam, Northern Ethiopia. Research Journal of Agriculture and Environmental Management, ISSN 2315-8719. Vol. 4(3), pp. 158-163.
25. UCI, "Abalone Data Set." https://archive.ics.uci.edu/ml/datasets/Abalone. Date Accessed: 9th January 2017.
26. Shapire R, Freund Y , Bartlett P, Lee W. (1998) Boosting the margin: A new explanation for the effectiveness of voting methods. Annals of Statistics, Vol. 26, No. 5, pp. 1651-1686.
27. Gagne SG. (2005). Operating Systems Concepts, 7th Edition. John Wiley and Sons, 2005.

28. Anitha A, Acharyjya DP. (2017). Crop suitability prediction in Vellore District using rough set on fuzzy approximation spapce and neural network. The Natural Computing Applications Forum 2017, Springer.
29. Hamzeh S, Mokarram M, Haratian A, Bartholomeus H, Ligtenberg A, Bregt AK. (2016). Feature selection as a time and cost-saving approach for land suitability classification (Case Study of Shavur Plain, Iran). Journal of Agriculture, Multidisciplinary Digital Publishing Institute, Vol. 6, Issue No. 4.
30. Mokarram M, Hamzeh S, Aminzadeh F, Zarei AR. (2015). Using machine learning land suitability classification. West African Journal of Applied Ecology. Vol. 23, Issue No. 1, pp. 63-73.
31. Dahikar SS, Rode SV. (2014). Agricultural crop yield prediction using artificial neural network approach. International Journal of Innovative Research in Electrical, Electronics, Instrumentation and Control Engineering, Vol. 2, Issue No. 1.
32. Elsheikh R, Shariff ARBM, Amiri F, Ahmad NB, Balasundram SK, Soom MAM. (2013). Agriculture Land Suitability Evaluator (ALSE): A decision and planning support tool for tropical and subtropical crops. Journal of Computers and Electronics in Agriculture. Elsevier Vol. 93, pp. 98-110.
33. Kotsiantis SB, Zaharakis ID, Pintelas PE. (2016). Machine learning: a review of classification and combining techniques. Artificial Intelligence Review, Vol. 26, Issue No. 3, pp. 159-190, Springer.

Classification of Brain Activity Using Brainwaves Recorded by Consumer-Grade EEG Devices

Elizaveta Kosiachenko, Dong Si

Computing and Software Systems, University of Washington Bothell, USA

(elizavet@uw.edu, dongsi@uw.edu)

Abstract. Despite the significant amount of research being conducted in interpreting electroencephalography (EEG) brainwave recordings using machine learning techniques, most of it is done using multi-electrode, medical grade EEG systems. However, these devices are expensive and hard to use by non-professional users. Our research focuses on analyzing the feasibility of using low-cost consumer grade EEG devices for classifying the type of activities performed by the user and for identifying a person based on their EEG measurements. This research is important as consumer grade EEG devices can potentially have a much wider application for the purposes of brain-computer interface than the lab alternatives. We have implemented k-nearest neighbor, decision tree and support vector machine algorithms in our analysis. The results show that consumer-grade EEG devices can be successfully used in some domains to both classify the type of activity performed by a person and to identify people based on their brainwave readings.

Keywords. Brain activity, consumer-grade non-invasive electroencephalography (EEG), classification, k-nearest neighbor, decision tree, support vector machine.

1. Introduction

Brain-computer interface is a very popular and fast-developing topic nowadays. One of the most common ways to implement such kind of interface is by using a non-invasive electroencephalography (EEG). There are quite a few research papers exploring the application of modern machine learning techniques to analyze and interpret EEG data: to perform emotional state classification [2], to identify depression patients [3], to control computer cursor [6] and even to use brain electrical activity as a biometric [5]. However, most of the research papers identified perform their experiments and evaluation using multi-electrode, medical grade EEG systems [3], [4], [5], [7], and only very few works focus on low-cost consumer grade EEG devices [8].

In the recent years a range of inexpensive and easy-to-use EEG devices became available, enabling taking this technology outside of the laboratory and bringing it into real-world environment. Considering the current availability of many different commercial consumer-grade EEG devices and their continuous development and update, there is a need in exploring the feasibility of using such low-cost EEG devices for monitoring individuals' EEG signals in their natural environment and making meaningful predictions based on this data.

The goal of this research article is to analyze how the consumer-grade EEG devices could be used for the following two tasks: 1) classifying the type of activity for a particular person based on their EEG reading; and 2) identifying a person based on their EEG measurements while doing a particular activity.

This research is based on the dataset generated by the BioSENSE group in the UC Berkeley using "Mindwave" consumer-grade EEG devices [1], which are produced by the Neurosky (http://neurosky.com/). The dataset was published on the UC Berkeley website (http://biosense.berkeley.edu/indra_mids_5_15_dlpage/).

2. Method & Approach

The high-level representation of the workflow of the experiments performed is as follows:

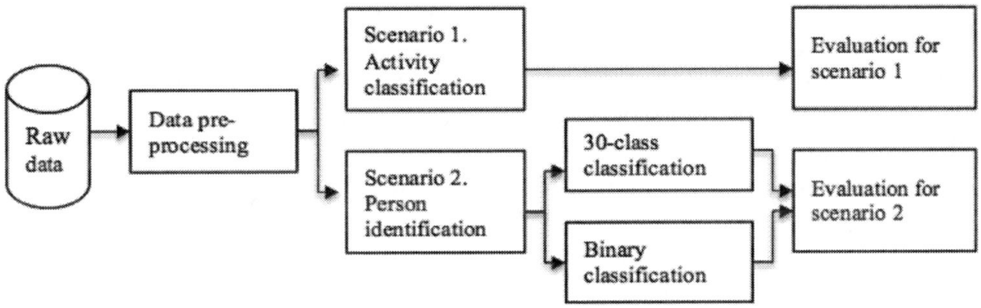

Fig. 1. Experiment workflow.

As mentioned above, the dataset analyzed in this paper is provided by the UC Berkeley researchers [1]. According to the description of the experiment, the data was collected using consumer-grade brainwave-sensing headsets "Mindwave" by Neurosky during an in-class group exercise. As part of the exercise participants were presented with various audio-visual stimulus. The dataset includes all subjects' readings during the stimulus presentation, as well as readings from before the start and after the end of the stimulus.

The types of exercises and stimulus presented to the participants are:

- blinking
- relaxing (with eyes closed)
- doing math
- listening to music (with eyes closed)
- watching video (TV commercial)
- thinking of items belonging to a particular category (with eyes closed)
- counting color blocks

The initial dataset contains 29,480 data items and 8 attributes. Each data item represents one reading of the brainwave activity. During the experiment the readings were collected every second from each "Mindwave" device. Data attributes collected during each reading are as follows: id, indra_time, browser_latency, reading_time, attention_esense, meditation_esense, eeg_power, raw_values, signal_quality, createdAt, updatedAt.

Data attributes represent the following information:

- **id**: Integer value in the range of 1 to 30 representing the participant number.
- **indra_time**: The synchronized timing of the reading.
- **browser_latency**: The difference between the time on the subject's computer, and the time on the server. This value is used to calculate the synchronized indra_time, above.
- **reading_time**: The time at which the Neurosky data passed through the bluetooth connection onto the subject's computer.
- **raw_values**: Tuple containing raw sample values acquired by the sensor, at a sampling rate of 512Hz (defined by the Neurosky SDK).
- **attention_esense** and **meditation_esense**: Neurosky's eSense meters for Attention and Meditation levels, in integer values in the range of 0 to 100. The values represent the last values computed in the current period (defined by the Neurosky SDK).
- **eeg_power**: Tuple represents the magnitude of 8 commonly-recognized types of EEG frequency bands -- delta (0.5 - 2.75Hz), theta (3.5 - 6.75Hz), low-alpha (7.5 - 9.25Hz), high-alpha (10 - 11.75Hz), low-beta (13 - 16.75Hz), high-beta (18 - 29.75Hz), low-gamma (31 - 39.75Hz), and mid-gamma (41 - 49.75Hz). These values have no units (defined by the Neurosky SDK).
- **signal_quality**: A zero value indicates good signal quality. A value of 128 greater corresponds to a situation where the headset is not being worn properly (defined by the Neurosky SDK).

Brainwave data is characterized by significant subject-to-subject variability [5], so it was reasonable to perform the classification by the type of activity only in the context of each particular person. We focused on data pre-processing and prediction on two scenarios:

1. Classifying the type of activity for a particular person based on their EEG reading;
2. Identifying a person based on their EEG measurements while doing a particular activity.

Both scenarios outlined above are supervised learning problems, various machine learning algorithms were implemented for this classification task. All pre-processing and classification was performed using Python programming language and a set of libraries for data manipulation and machine learning ("numpy", "pandas", "sklearn", "re", and "pylab").

2.1. Data Pre-processing

To prepare the data for training classification models and clear it from noisy readings, the following steps were performed:

- Removed unused columns (data attributes), such as indra_time, browser_latency, reading_time (because activity labels are already assigned to the data readings, time information is irrelevant);
- Removed the "raw_values" column, as, according to the manufacturer, the data from the "raw_values" column is aggregated into 8 EEG bands represented in the "eeg_power" column used in this research;
- Removed the "attention_esense" and "meditation_esense", as these attributes represent the emotional state of the subject and our research aim to classify between tasks performed by the subjects;
- Split "eeg_power" into 8 columns, which represent 8 commonly-recognized types of EEG frequency bands (delta (0.5 – 2.75Hz), theta (3.5 - 6.75Hz), low-alpha (7.5 - 9.25Hz), high-alpha (10 - 11.75Hz), low-beta (13 - 16.75Hz), high-beta (18 - 29.75Hz), low-gamma (31 - 39.75Hz), and mid-gamma (41 - 49.75Hz));
- Filtered out all the entries with "signal quality" > 128, which represent the readings taken when the signal quality was poor or when the device wasn't worn properly;
- Deleted all entries with labels "unlabeled" and "instructionsForX";
- Renamed labels like "blink1", "blink2", *etc.* into just "blink".

2.2. Methods & Approach for Scenario One – Activity Classification

The aim of the first scenario was classifying the type of activity for a particular person based on their EEG reading.

For the purposes of analyzing this scenario, we selected two distinct activities out of those performed by the subjects of the original research: relaxation with eyes closed and color round (*i.e.*, counting color blocks displayed on the screen). So, the first step in this scenario was to filter out only the readings with the values "relax" and "colorRound" in the column "label".

Then the following steps were performed for each of the participants in the dataset. First, we filtered out other readings and kept the readings related only to this participant. Second, we trained the k-nearest neighbor (kNN)[1], decision tree[2] and support

---

[1] In k-nearest neighbor(KNN) classification, an object is classified by a majority vote of its neighbors, with the object being assigned to the class most common among its k nearest neighbors. Distance-weighted k-nearest neighbor classifier also uses a distance function to weight the evidence of a neighbor depending on their distance from an unclassified observation [9].

[2] Decision trees are a supervised learning method, which maps observations about an item (represented in the branches) to conclusions about the item's target value (represented in the leaves). The goal is to create a model that predicts the value of a target variable by learning simple decision rules inferred from the data features [10].

vector machine (SVM)[3] with linear and polynomial kernel models. Finally, the 5-fold cross-validation was performed to assess the quality of model prediction.

For the k-nearest neighbor model we conducted a series of experiments with different numbers of k and selected the one that gave the best accuracy of the prediction (k=25). We also experimented with the different degrees of the polynomial kernel function and the best result for this problem was produced by the degree of 3.

Because the number of data items in the two classes selected ("relax" and "colorRound") is not equal, we used the stratified 5-fold cross-validation as a method of model assessments.

2.4. Methods & Approach for Scenario Two – Person Identification

The aim of the second scenario was to predict the person identification based on their EEG measurements while doing a particular activity. Two different experiments - 30-class classification and binary classification - were designed to achieve that goal.

30-class classification. In this experiment, each class represents readings for a particular person. We selected one of the activities' labels ("math") and filtered out all of the readings with other labels. We then trained the following models to classify the EEG readings between 30 "id" labels: k-nearest neighbor (KNN), decision tree and SVM with linear and polynomial kernels. A 5-fold cross-validation was then performed to assess the accuracy of each model.

Binary classification. Because 30-class classification provided poor prediction accuracy (for details see the section 3.2. Results of Scenario Two), we designed an alternative experiment in hopes of getting better prediction results. For each of the participants and most of the activities in the dataset (excluding blinking activity due to low number of data readings belonging to this class), the following steps were performed:

- Create a new dataset and only copy the readings related to the activity that is currently analyzed from the full pre-processed dataset;
- Substitute all the "id" labels for the readings related to the participant in question with '1's;
- Substitute all other participants "id" labels with '0' and randomly delete rows from this category until the number of rows with '0' in the 'id' column equals the number of rows with '1's;
- Use the resulting data subset to: :
  - Train the k-nearest neighbor (KNN), decision tree and SVM with linear and polynomial kernel models to classify between the two classes = [0, 1];
  - Assess the accuracy using stratified 5-fold cross-validation.

We also performed a grid search for the best k in the k-nearest neighbor algorithm separately for each of the activities, but across subjects. We also tried various degrees

[3] Support Vector Machines (SVMs) are hyperplanes that separate the training data by a maximal margin. All vectors lying on one side of the hyperplane are labeled as one class, and all vectors lying on the other side are labeled as another class. SVMs allow to project the original training data in space to a higher dimensional feature space via a kernel operator [11].

of the polynomial kernel in the SVM model and noticed that the best prediction for this scenario was achieved by using the quadratic kernel (degree of 2).

3. Experiment Results

3.1. Results of Scenario One

The results of running the abovementioned models (KNN, Decision tree, SVM – linear kernel and SVM – polynomial kernel) for each of the participants are presented in the table below. The metric used for assessing the quality of prediction is accuracy, which for this scenario is defined as the sum of correctly predicted readings from the "relax" class and correctly predicted readings from the "colorRound" class divided by the total number of observations from both classes.

Table 1. Accuracy of models predictions for Scenario One.

	KNN	**Decision Tree**	**SVM (linear)**	**SVM (polyn.)**
Subject 1	73.37%	62.46%	72.46%	72.46%
Subject 2	72.46%	67.79%	72.46%	72.46%
Subject 3	77.10%	74.24%	75.27%	75.27%
Subject 4	73.39%	63.29%	76.20%	77.03%
Subject 5	84.48%	72.39%	82.57%	83.36%
Subject 6	72.46%	55.93%	72.46%	72.46%
Subject 7	76.95%	74.63%	76.20%	70.66%
Subject 8	72.90%	67.09%	72.90%	71.94%
Subject 9	72.46%	60.51%	72.46%	72.46%
Subject 10	73.11%	77.70%	77.87%	78.78%
Subject 11	76.12%	76.99%	81.66%	78.06%
Subject 12	70.38%	66.66%	70.38%	70.30%
Subject 13	91.73%	87.22%	91.68%	92.72%
Subject 14	74.15%	64.02%	75.02%	72.25%
Subject 15	71.14%	58.76%	72.14%	72.14%
Subject 16	73.85%	64.54%	73.76%	70.04%
Subject 17	71.31%	54.63%	71.31%	71.31%
Subject 18	69.22%	60.56%	71.94%	71.94%
Subject 19	78.85%	68.58%	71.31%	71.31%
Subject 20	71.90%	74.80%	74.67%	73.85%
Subject 21	80.43%	72.77%	82.20%	79.43%
Subject 22	71.94%	73.93%	76.58%	73.85%
Subject 23	72.85%	68.22%	72.90%	71.94%
Subject 24	71.90%	66.23%	70.04%	71.94%

Subject 25	71.14%	50.00%	71.14%	71.14%
Subject 26	72.26%	61.34%	72.31%	73.09%
Subject 27	75.75%	86.01%	84.02%	74.80%
Subject 28	71.94%	60.77%	71.94%	71.94%
Subject 29	93.42%	92.53%	78.98%	76.65%
Subject 30	71.57%	64.21%	74.30%	70.66%
Average accuracy of the model among the participants	**75.02%**	**68.30%**	**75.31%**	**74.21%**

As shown in the Table **1**, the best average accuracy of prediction of 75.31% is provided by the Support Vector Machine model with the linear kernel. However, there is quite a large variance in the quality of prediction between the participants. For example, for Subject 29 the best model has an 93.42% accuracy of prediction and for Subject 28 the best prediction accuracy is only 71.94%. One of the explanations for such a variance is that some participants might have followed the instructions and focused on the exercise better then the others.

In general, the prediction accuracy of 75.31% is a result, which can provide some useful information (for example, it can be used in a gaming environment where incorrect prediction is not critical), however this accuracy is not good enough for such predictions to be the basis of some important decisions.

3.2. Results of Scenario Two

As mentioned above, for the purposes of Scenario Two we designed two different experiments: 30-class classification and binary classification.

30-class classification result. Unfortunately, after training the models (KNN, Decision tree, SVM – linear kernel, SVM – polynomial kernel) to classify the EEG readings between 30 ‘id’ labels, the accuracy of prediction provided by these models was pretty low:

Table 2. Accuracy of models predictions for Scenario Two – 30-class classification.

Model	Prediction accuracy (using 5-fold cross-validation)
K-nearest neighbor (k = 4)	25.10%
Decision tree	22.45%
SVM – linear kernel	22.55%
SVM – polynomial(quadratic) kernel	15.05%

The accuracy of 25.10%, demonstrated by the k-nearest neighbor model, although is better than the random guess (1/30 = 3.33%), can not realistically be used to solve any real-world problems.

Binary classification. In the hopes of getting a better performance than the 30-class classification we designed another experiment to classify brainwave readings as belonging to a certain individual or not – binary classification (see the details of algorithm steps in section 2.3. Methods & Approach for Scenario Two). The accuracy, precision and recall of different models trained on the data from different activities are presented in the figures below.

For this scenario accuracy is defined as the sum of readings correctly predicted as belonging to the individual and correctly predicted as not belonging to the individual, divided by the total number of readings from both of these classes. We define precision to be the number of readings correctly predicted as belonging to the individual divided by the total number of readings predicted as belonging to the individual. We define recall to be the number of readings correctly predicted as belonging to the individual divided by the total number of readings actually belonging to the individual being analyzed.

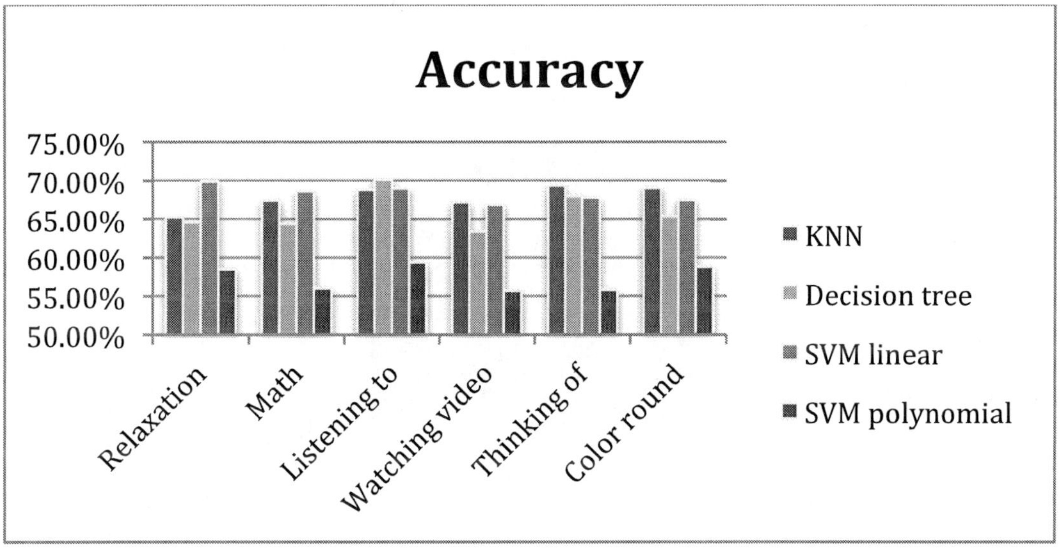

Fig. 2. Average accuracy of models for Scenario Two – binary classification.

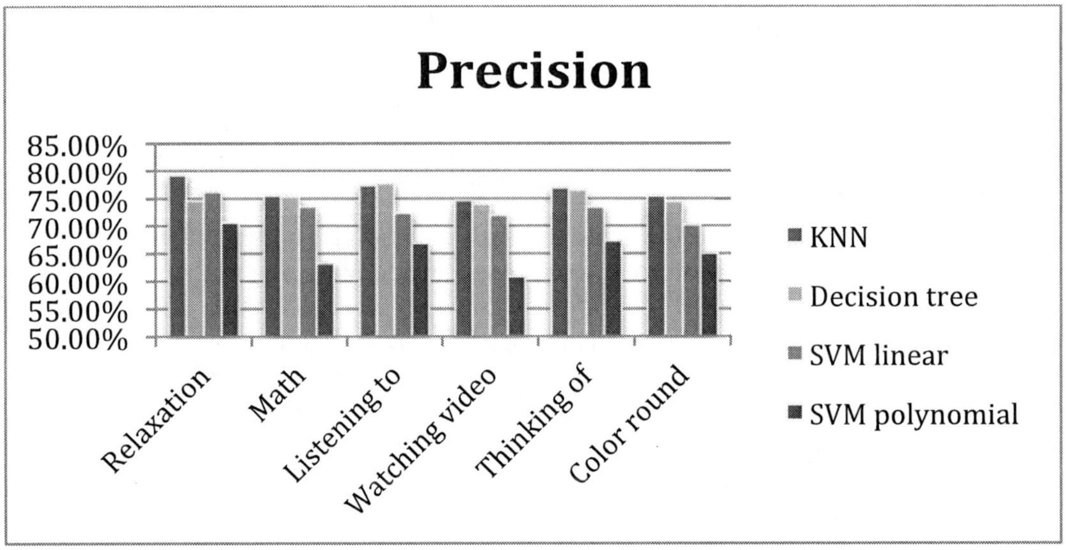

Fig. 3. Average precision of models for Scenario Two – binary classification.

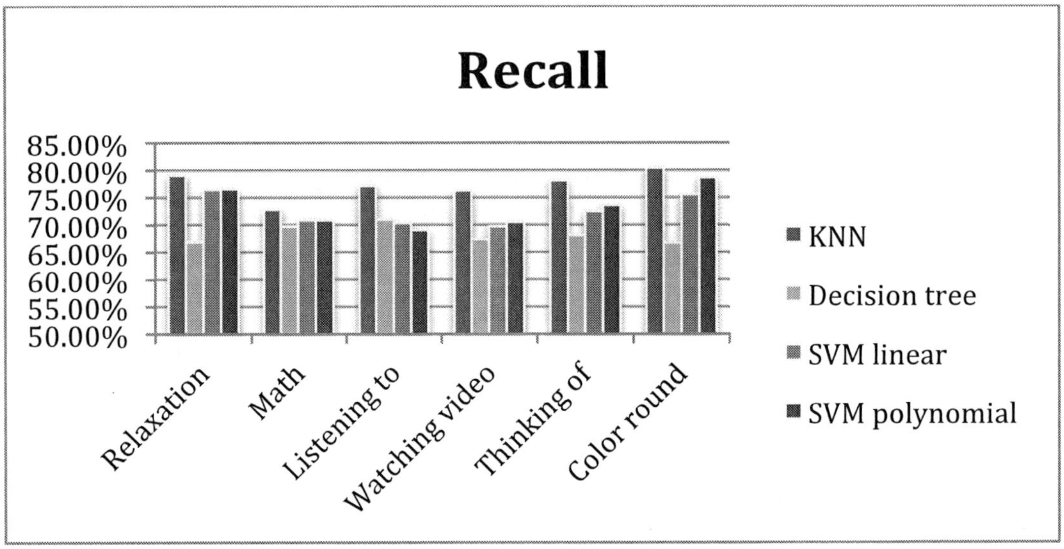

Fig. 4. Average recall of models for Scenario Two – binary classification.

As we can see from the Figures **2** to **4**, binary classification does provide better results than 30-class classification and proves that consumer-grade EEG devices work. The best accuracy of prediction is provided by applying the decision tree algorithm to the brainwave data collected while participants were listening to music and equals 70.18%. Curiously, the worst activity for identifying people, based on the results of our experiment, is watching video (in the original experiment participants were watching TV commercial). The best accuracy of prediction for the activity of watching video is only 67.23% and is achieved using k-nearest neighbor algorithm with k=8.

However, the nature of the applications that might use person identification from the brainwave recordings would probably require a very high accuracy of prediction, which so far is not achieved by the methods described.

There are two factors that might negatively influence the accuracy of prediction. The first is a relatively small dataset for the purpose of this scenario. We noticed that, despite the fact that the original dataset contained about 30,000 reading, after clearing the data, filtering out irrelevant data and splitting readings by participants and by activities, the amount of data per participant per activity was only about 30 readings, which is not ideal for making highly accurate and precise predictions. And the second factor could be the imperfections of the underlying technology used in the consumer-grade EEG devices used in the research.

4. Conclusion and Discussion

Based on the results from Scenario One, the data from consumer-grade EEG devices produced by the Neurosky, can be successfully used to classify the type of activity for a particular person based on their EEG reading. The results vary greatly from person to person, but the average accuracy of 75.31% allows this prediction to be used in some potential domains, such as gaming or other entertainment.

Scenario Two had a slightly lower quality of prediction with the best average accuracy provided by the 'music' activity at 70.18% and the worst average accuracy provided by the 'video' activity at 67.23%. This result can still be used in some applications and proves that the consumer EEG devices are capable of distinguishing between the brainwaves of different people. However, the nature of applications that use person identification requires a very high accuracy of prediction, which is not delivered by the dataset at hand. So the application of the results of the second experiment is quite limited.

We expect the current fairly good accuracy of prediction could be further improved in the future when the technology used in the consumer-grade EEG devices is improved and more high quality data are collected.

The commercial-grade EEG devices are still a developing technology that needs improvement in the quality of brainwave identification and recording in order to be used to successfully classify brainwaves by activity or to be used to identify people based on their EEG readings. However, these devices are developing really fast, so it is promising that this technology will have a big potential.

For the future work, we would like to continue working on the two scenarios outlined in this paper, in particular to look into improving models' accuracies as new generations of consumer-grade EEG devices are being released. After more and more good quality data has been generated from the devices, some advanced prediction models such as Artificial Neural Networks could be implemented. We think it will be a very interesting and promising concept with wide array of potential applications in brain-computer interface domain.

References

1. Chuang J, Merrill N, Maillart T. (2015). Synchronized Brainwave Recordings from a Group Presented with a Common Audio-Visual Stimulus. UC Berkeley Spring 2015 MIDS Immersion Class.
2. Wang X, Nie D, Lu B. (2014). Emotional state classification from EEG data using machine learning approach. Neurocomputing, 129, 94-106.
3. Hosseinifard B, Moradi M, Rostami R. (2013).Classifying depression patients and normal subjects using machine learning techniques and nonlinear features from EEG signal. Computer Methods and Programs in Biomedicine, 109, 339-345.
4. Blankertz B, Dornhege G, Krauledat M, Muller K, Kunzmann V, Losch F, Curio G. (2006). The Berlin brain–computer interface: EEG-based communication without subject training. IEEE Transactions on Neural Systems and Rehabilitation Engineering, 14, 147-152.
5. Palaniappan R, Mandic D. (2007). Biometrics from Brain Electrical Activity: A Machine Learning Approach. IEEE Transactions on Pattern Analysis and Machine Intelligence, 29, 738-742.
6. Kostov A, Polak M. (2000). Parallel man-machine training in development of EEG-based cursor control. IEEE Transactions on Rehabilitation Engineering, 8, 203-205.
7. Müller K, Tangermann M, Dornhege G, Krauledat M, Curio G, Blankertz B. (2008). Machine learning for real-time single-trial EEG-analysis: From brain–computer interfacing to mental state monitoring. Journal of Neuroscience Methods, 167, 82-90.
8. Maskeliunas R, Damasevicius R, Martisius I, Vasiljevas M. (2016). Consumer grade EEG devices: are they usable for control tasks?. Peer J, 4, e1746.
9. Dudani S. (1976). The distance-weighted k-nearest-neighbor rule. IEEE Transactions on Systems, Man, and Cybernetics, SMC-6, 325-327.
10. Quinlan J. (1985). Induction of decision trees. New South Wales Institute of Technology, School of Computing Sciences, [Broadway, N.S.W., Australia].
11. Abe S. (2005). Support vector machines for pattern classification. Springer, London.

A Three-dimensional Coordinate Positioning Method of Fish Target Based on Computer Vision Using A Single Camera

Ase Shen[1], Gang Xiao[1,*] , Jiafa Mao[1], Haibiao Hu[1], Weiguo Sheng[1], Li-Nan Zhu[1]

[1] College of Computer Science & Technology, ZheJiang University of Technology, Hang Zhou, 310023, China
[*] Corresponding author. E-mail: xg@zjut.edu.cn

Abstract. In order to improve the analysis efficiency of fish behavior in the three-dimensional space, and provide key technologies for the water quality monitoring, this paper proposed a method to obtain the three-dimensional coordinate positioning of the fish using a single camera based on computer vision technologies. According to the principle of the underwater object imaging, camera imaging principle, object segmentation, plane geometry and image processing techniques, this paper discussed the two-dimensional image object extraction process, and deduced the theoretical model of the three-dimensional coordinate positioning reconstructed based on the two-dimensional image. The effectiveness and accuracy of the proposed method were demonstrated by the experimental results of live fish and artificial fish on the system platform.

Keywords: Plane mirror imaging, underwater object imaging, three-dimensional positioning, target segmentation.

1. Introduction

The increasingly serious water pollution is a source of disease which threatens the safety of human life. It is a serious obstacle to human health, economic development and social stability. How to carry on the reliable monitoring and timely warning to the water quality is the key link in the water pollution prevention and control. The traditional water quality testing technology mainly uses the biological sampling inspection method, which is difficult to reflect the water quality status in real time. Fish are very sensitive to changes of water environment, especially toxic substances such as pollutants. When the water is contaminated, the respiratory rate of fish, gill cover activity rate, speed, gregariousness, magnetism and other behavior will change [1]. Therefore, it is possible to capture the normal and abnormal behavior of fish by computer vision and realize real-time observation, inspection and analysis of water pollution to achieve the real-time monitoring of water quality.

Fish have acceleration, avoidance, mirror bite, migratory, wagging tail and other acts. In most of the current experimental environment, the extraction of fish motion parameters is based on video recording, and the extraction of fish behavior from video has been a popular research topic [2]. In the three-dimensional space, fish behavior data parameters are swimming speed, trajectory, fish location, *etc.* And to carry out parameter analysis, the accurate three-dimensional positioning of fish should be determined [3]. Normally, the three-dimensional positioning of fish is restored by two video frames form two cameras in different angles. Visdcido [4] collected three-dimensional information of fish via two cameras, one on the top of the water tank and the other in front. Kittiwann [5] placed cameras in the front of the water tank and in the right side of the tank to record the swimming situation of zebrafish in the XZ plane and the YZ plane, and then used these two records to restore the zebrafish's three-dimensional space information. Wu Guanhao [6] shot the bottom view and side view of the fish tank with two cameras to achieve three-dimensional fish tracking measurement. Although these methods can locate the fish in three-dimensional coordinates, the amount of video data form dual cameras is large, so the real-time algorithm is difficult to achieve. Pereira [7] used a water-proof mirror placed on the left side of the water tank at 45°, and used a camera to record the fish target information among the fish tank and the waterproof mirror to restore the three-dimensional coordinate of the fish target. Derry [8] proposed to put the waterproof mirror on the right side of the water tank at 45°, with a single camera for data collection, and restore the three-dimensional information of organisms in the tank. Xu Panlin [9] achieved a single camera three-dimensional imaging by putting the water mirror placed in the top of the fish tank. Zhu [10] proposed a monocular three-dimensional detection system, which used two mirrors and a single camera to achieve the monocular three-dimensional information measurement.

Based on the previous research, this paper presents a method to realize three-dimensional coordinate positioning of fish target by single camera. Firstly, the three-dimensional data is collected by single camera and waterproof mirror. Then, according to the principle of plane imaging, the principle of underwater object imaging, the principle of camera imaging, the principle of plane geometry, the target segmentation and other image processing technology, two-dimensional image target is extracted. Finally, the three-dimensional position of the fish is calculated based on the two-dimensional coordinates information. The method proposed in this paper can effectively solve the problem of positioning error caused by refractive index due to the refractive index difference of different water quality. The experimental results show the effectiveness and accuracy of the proposed method of three-dimensional coordinate location.

This paper is organized as follows: the experimental platform design principles were introduced in Section 2. The principle of underwater objects imaging and the principle of camera imaging were described in Section 3. Section 4 described image segmentation and target location techniques. Section 5 was the positioning technology from two-dimensional coordinates to three-dimensional coordinates. The experimental results were shown in Section 6. The conclusion of the paper was offered Section 7 concludes.

2. The Design Principle of the System Platform

2.1. The System Platform Architecture

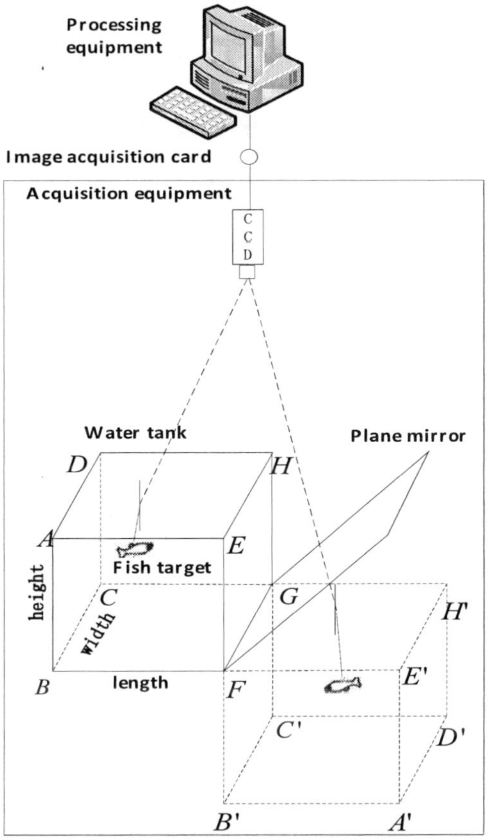

Fig. 1. target information acquisition system platform.

In order to realize the three-dimensional coordinate positioning by single camera, this paper designs a target information acquisition system platform. This system platform consists of three parts: acquisition equipment, image acquisition card, processing equipment. The acquisition equipment is mainly composed of a water tank and a plane mirror. The plane mirror is placed in the right side of the tank at a certain angle (angle of 45 °). The camera is placed directly above the water tank and mirror, used to collect the target information in the water tank and the plane mirror. Its shooting range covers the entire water tank and the tank image in the mirror. The processing equipment mainly carries out the processing of the collected image information. The entire acquisition system platform as shown in Figure **1**.

2.2. The Correspondence Between the Real Fish and its Mirror Image

According to the principle of plane mirror imaging and the principle of light reflection, the image in the plane mirror has the following corresponding relationship with the real object [11]:

1. The distances from the corresponding points of the mirror image and the object to the plane mirror are equal;
2. The mirror image is the same size as the object;
3. The connection line between a mirror image point and its corresponding point of the object is perpendicular to the plane mirror.

The correspondence between the real fish and its mirror image is shown in Figure **2**. Assuming that the three-dimensional coordinate of the fish in the tank is $S(X_S,Y_S,Z_S)$, the coordinate of its mirror image is $S''(X_{S''},Y_{S''},Z_{S''})$, by the above corresponding relationship 3 that can be easily obtained:

$$Y_S = Y_{S''} \ . \qquad \textbf{(1)}$$

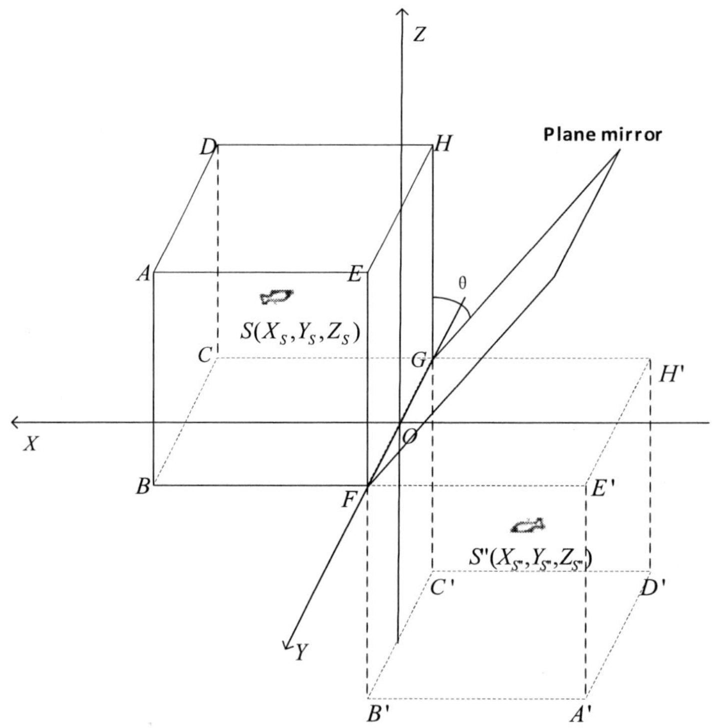

Fig. 2. The correspondence between the real fish and its mirror image.

Therefore, the coordinate relationship between the fish and its mirror image is converted into the coordinates (X,Z). As shown in Figure **3**, suppose that in the plane (X,O,Z), the fish coordinate is $S(X_S,Z_S)$ and its mirror image coordinate is $S''(X_{S''},Z_{S''})$.

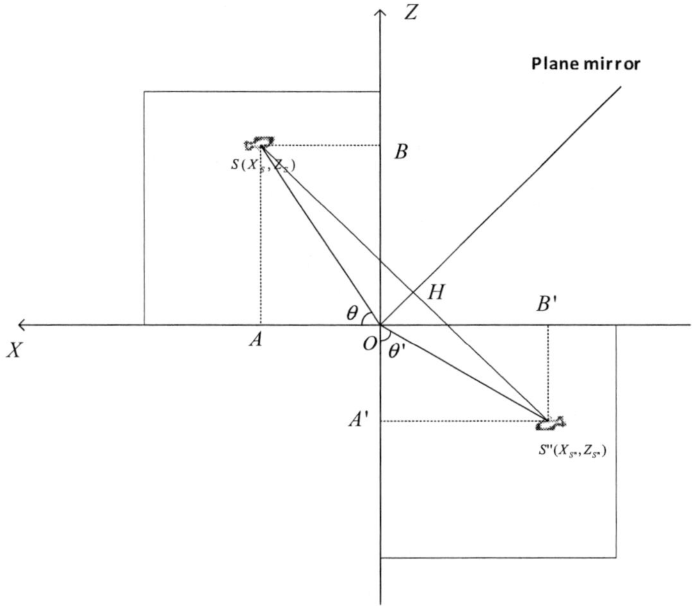

Fig. 3. The correspondence diagram of the fish in the water tank and its mirror image.

According to the plane mirror imaging principle and image similarity principle, the following correspondence can be obtained:

$$\begin{cases} \theta = \theta' \\ SO = S''O \end{cases} \quad (2)$$

Then, according to the conversion relationship between the hypotenuse and right angle side of the right triangle, we can get:

$$\begin{cases} Z_s = SA = SO\sin\theta \\ X_s = SB = SO\cos\theta \\ |Z_{S''}| = S''B' = S''O\cos\theta' \\ |X_{s''}| = S''A' = S''O\sin\theta' \end{cases} \quad (3)$$

From equation (2) and equation (3), this can be solved:

$$\begin{cases} Z_s = -X_{S''} \\ X_S = -Z_{S''} \end{cases} \quad (4)$$

Therefore, with the equation (1) and equation (4), the corresponding relationship between the fish in the water tank and its mirror image in the plane is:

$$\begin{cases} X_S = -Z_{S''} \\ Y_S = Y_{S''} \\ Z_S = -X_{S''} \end{cases} \quad (5)$$

3. The Principle of Underwater Object Imaging

3.1. Underwater Object Imaging Process

Assuming that the underwater object M is at the point O, and when the object M emits light at the point O, the light is refracted on the water surface, but not all of the refracted rays intersect at the same point. As shown in Figure **4**, the reverse extension lines of AA' and BB' intersect at S_1, the reverse extension lines of BB' and CC' intersect at S_2, S_1 and S_2 does not overlap.

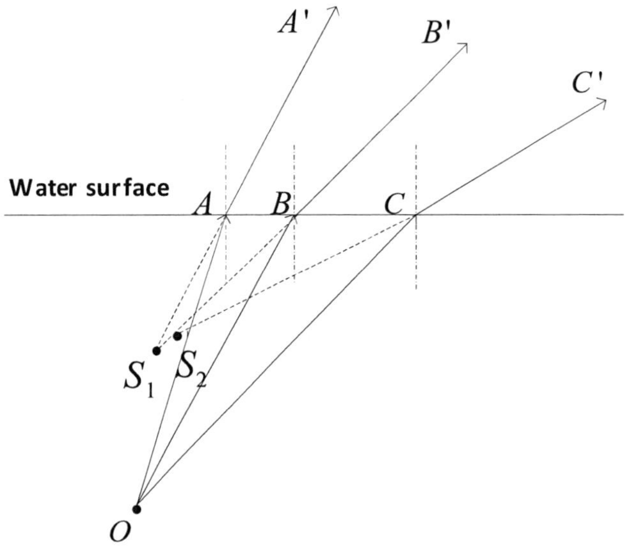

Fig. 4. The light path of underwater object imaging.

When the light emitted by the O point is very narrow, the refracted light should also be also very narrow, and the intersection of their reverse extension line can be approximated as one point to form an approximate image. Therefore, when the observation aperture is small enough, and the lights entering the observation aperture are also very narrow, the reverse extension lines of the refracted lights will be approximately intersected at one point, and the image of the object in the water can be observed at the observation point [12,13].

3.2. Imaging Process

Based on the information acquisition platform designed in this paper, the imaging process of fish target is shown in Figure **5**. Suppose the point S is the actual position of the fish in the three-dimensional space, and the point S emits two light ray, line SA and line SB , after the water surface refraction, the reflected lights are AA' and BB', whose reverse extension lines intersect at point S'. Assuming that refracted light AA' can just enter the camera, when A' is infinitely close to B', according to the imaging principle of underwater object and the principle of pinhole imaging, the intersection point *S'* can be approximately taken as the location of the virtual image.

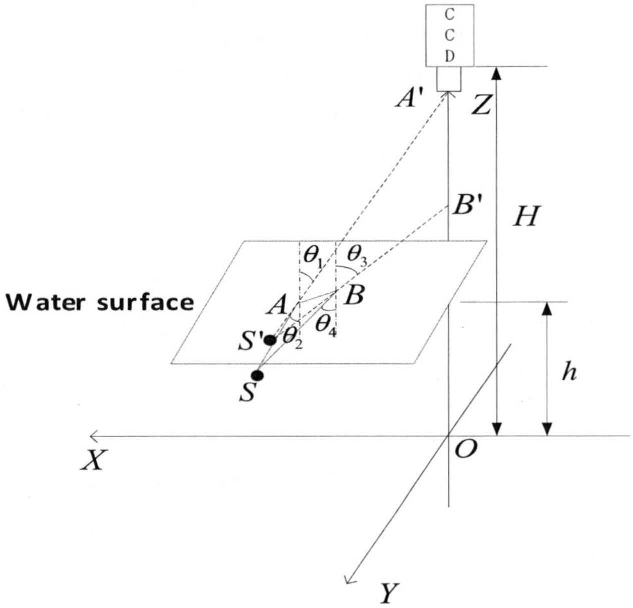

Fig. 5. The diagram of underwater object imaging in the information acquisition platform.

Suppose the height of the camera from the tank bottom is H (cm), the height of the water surface in the tank is h (cm), the coordinates of the fish target is $S(X_S, Y_S, Z_S)$ and the coordinate of its virtual image is $S'(X_{S'}, Y_{S'}, Z_{S'})$. *S'* on the reverse extension lines of *AA'* and *BB'*, the coordinate of A is $A(X_A, Y_A, Z_A)$,the coordinate of *B* is $B(X_B, Y_B, Z_B)$. Because A and B are on the water surface, the values of Z_A, Z_B and h (cm) are equal.

When the light SA is refracted on the water surface, assuming the angle of incidence is θ_2 and the angle of refraction is θ_1, according to the refractive index calculation formula,, the refractive index of water *n* can be expressed as:

$$n = \frac{\sin\theta_1}{\sin\theta_2} \tag{6}$$

According to the coordinates in Figure **5**, the plane geometry and trigonometric functions, $\sin\theta_1$ and $\sin\theta_2$ can be expressed as:

$$\begin{cases} \sin\theta_1 = \dfrac{\sqrt{X_A^2 + Y_A^2}}{\sqrt{(H - Z_A)^2 + X_A^2 + Y_A^2}} \\ \sin\theta_2 = \dfrac{\sqrt{(X_S - X_A)^2 + (Y_S - Y_A)^2}}{\sqrt{(X_S - X_A)^2 + (Y_S - Y_A)^2 + (Z_S - Z_A)^2}} \end{cases} \tag{7}$$

According to the principle of pinhole imaging, the image S' captured by the camera is the virtual image of the fish target. Figure **5** shows the coordinates relationship between the virtual image S' and the fish target S. According to spatial analytic geometric transformation calculation principle, optical principle and the relationship of refractive index, the following relations between virtual image coordinates and actual fish coordinates can be obtained:

$$\begin{cases} X_{S'} = \dfrac{X_S}{\sqrt{X_S^2+Y_S^2}}\left(\sqrt{X_S^2+Y_S^2} - \dfrac{(n^2-1)(h-Z_S)\sin^3\theta_1}{(n^2-\sin^2\theta_1)^{3/2}}\right) \\ Y_{S'} = \dfrac{Y_S}{\sqrt{X_S^2+Y_S^2}}\left(\sqrt{X_S^2+Y_S^2} - \dfrac{(n^2-1)(h-Z_S)\sin^3\theta_1}{(n^2-\sin^2\theta_1)^{3/2}}\right) \\ Z_{S'} = h - \dfrac{n^2(h-Z_S)\cos^3\theta_1}{(n^2-\sin^2\theta_1)^{3/2}} \end{cases} \quad \textbf{(8)}$$

θ_1 is the refraction angle of the light SA, according to equation (7) and the plane trigonometric function conversion relationship, this can be obtained:

$$\begin{cases} \sin\theta_1 = \dfrac{\sqrt{X_A^2+Y_A^2}}{\sqrt{(H-h)^2+X_A^2+Y_A^2}} \\ \cos\theta_1 = \dfrac{H-h}{\sqrt{(H-h)^2+X_A^2+Y_A^2}} \end{cases} \quad \textbf{(9)}$$

4. Two-dimensional Image Target Positioning Algorithm

4.1. Image Segmentation

At present, color image segmentation techniques are divided into six types [14]: pixel-based, edge-based, region-based, model-based, physics-based and hybrid. Because the red crucian carp is used as the fish target, this paper adopts a color image segmentation method based on RGB color space [15-18] to segment the acquired video frames. The specific algorithm steps are as follows:

Step 1: Extract the current captured video frame image $f(R,G,B)$.

Step 2: Calculate the values of R, G and B of the currently extracted video frame, denoted by $f_R(R,G,B)$, $f_B(R,G,B)$ and $f_G(R,G,B)$. Then select the combination operators $(R-B)$, $(R-G)$, $(G-B)$ and $(R-G-B)$ as feature quantity to carry out algebraic operations and to perform color feature analysis and extraction in RGB space. And extraction. The results show that there is a significant difference between the gray value of $(R-B)$ and the gray value of the background, and the difference basically presents a bimodal distribution. Therefore, use the color difference feature of $(R-B)$ as the feature point of image segmentation, and using one-dimensional features can improve the segmentation efficiency.

Step 3: In the color space RGB, use the $(R-B)$ chromatic aberration gray value as the feature threshold to segment the image. Calculate the difference between $f_R(R,G,B)$ and $f_B(R,G,B)$ to obtain the gray value distribution of the image. According to the image chromatic aberration gray value, determine the threshold, then use the equation (10) to segment the image object:

$$|f_R(R,G,B)-f_B(R,G,B)|\geq threshold \tag{10}$$

Step 4: According to the equation (10), segment the image target and get the pixels that meet the condition. Figure (**6a**) is the original video frame and Figure (**6b**) is the result of the above segmentation algorithm results.

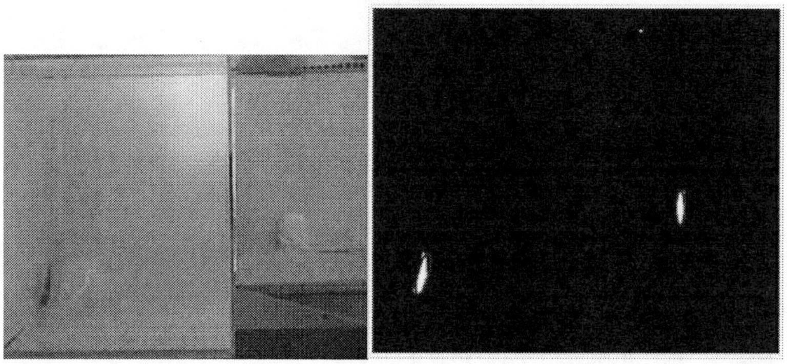

Fig. 6. The left is a original video frame. The right is a binary image after segmentation.

4.2. Target Location

As shown in Figure **6**, for the sake of convenience of expression, we express the fish in the left part of the video frame is expressed as real fish, which is the image of fish in the camera. The fish in the right part of the video frame is expressed as virtual fish, which is the image of the fish mirror image in the camera image. In Figure (**6a**), on the left side is the real fish, on the right is the virtual fish. After accurately detecting and segmenting the target image, the next step is to locate the target point accurately. Since the target is still in two-dimensional space and is composed of many pixels, how to select a point to represent the target has become an important factor affecting the accuracy of three-dimensional positioning. Because fish are constantly swimming and flipping, even in the same position, the coordinates of fish's gravity center are different. So it is not suitable for fish target location. In view of the information collection platform used in this paper, as well as the particularity of the selected object, this paper selects the central point as the location point of the fish target. As shown in Figure (**6b**), in the left side, the central point of the real fish is biased to the upper part of the fish body. In the right side is the virtual fish, its central point is also biased to the upper part. The two central points correspond exactly. Therefore, compared with the gravity center, the central point is more representative, more suitable for the experiment.

5. Three-dimensional Coordinate Positioning

5.1. Camera Imaging Model

The corresponding relationship between a point on the space and its imaging point is determined by the camera imaging model. The general camera imaging model is the pinhole camera model [19-22]. Figure **7** shows the camera imaging model in this experiment. Figure **8** gives the information acquisition platform imaging diagram.

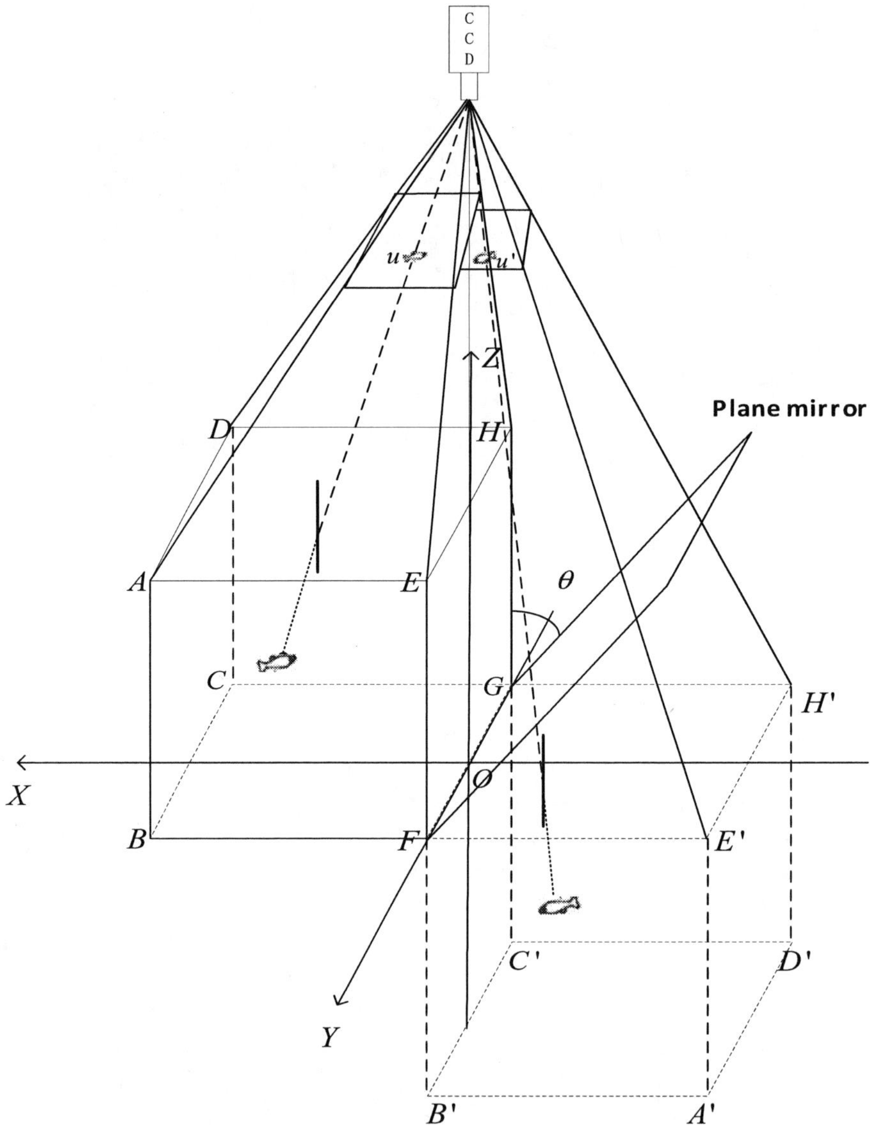

Fig. 7. The camera imaging model in this experiment.

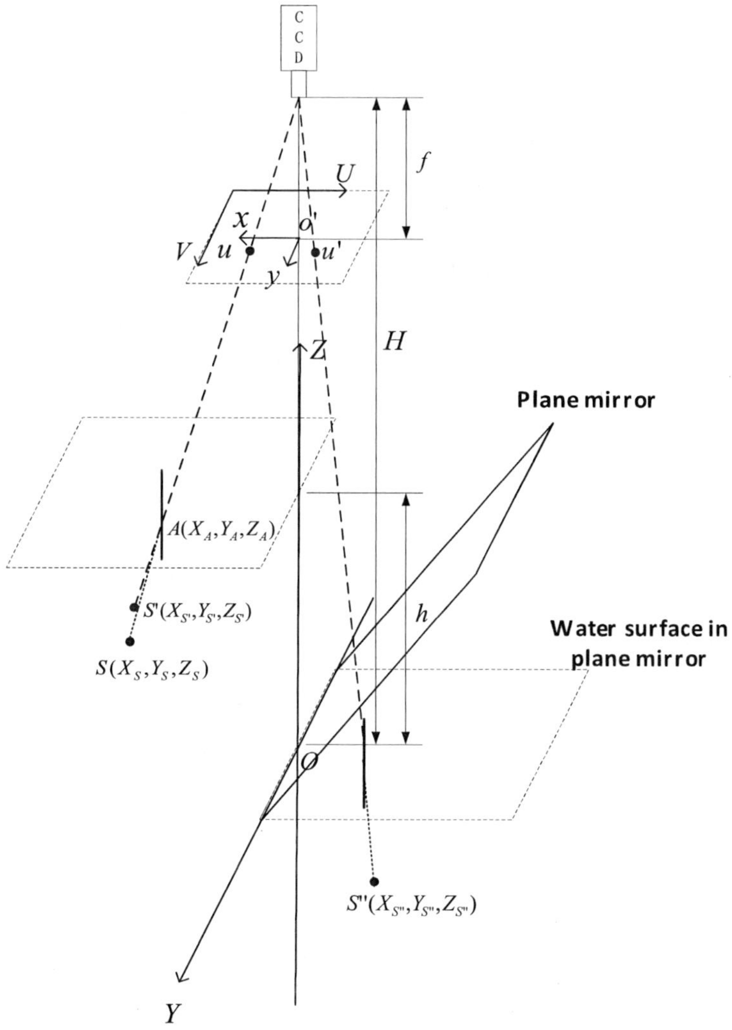

Fig. 8. The information acquisition platform imaging diagram.

In this experiment, the camera position was fixed above the origin of the coordinate axis, (X,O,Y), and the imaged image can cover the water tank on the left side and its virtual image in the mirror. The shooting point is at the middle point of the image.

Therefore, according to the mirror symmetry of the image and the similarity of the graph, the image size and the size of the actual object have the following correspondence:

$$\begin{cases} L' = 2L * \dfrac{f}{H - h_y} \\ W' = W * \dfrac{f}{H - h_y} \end{cases} \tag{11}$$

Where L' (cm) and W' (cm) are the length and width of the water tank on the captured image, respectively, L (cm) and W (cm) are the length and width of the actual water tank, respectively, f (cm) is the focal length of the camera, H (cm) is the height of the shooting point to the tank bottom, and h_y (cm) is the height of the actual tank.

As shown in Figure **8**, (X,Y,Z) is the coordinate system of the actual reference, that is, the absolute coordinate system. (U,V) is the image pixel coordinate system in pixels, and (x,o',y) is the image physical coordinate system in centimeters[23]. Assuming that d_x and d_y represent the pixel length of each pixel in the direction x and the direction y respectively, the transition relation between the position of the point on the image pixel coordinate system and the position on the image physical coordinate system is:

$$\begin{cases} x = d_x(u_0 - u) \\ y = d_x(v - v_0) \end{cases} \quad \text{(12)}$$

Where (u_0, v_0) is the coordinate of the origin of the image physical coordinate system in the image pixel coordinate system and (u,v) is the coordinate of the point in the image pixel coordinate system.

According to the length and width of the water tank on the captured image (L'(cm)and W'(cm)), d_y and d_y can be obtained by the following equation:

$$\begin{cases} d_x = \dfrac{L'}{u_t} \\ d_y = \dfrac{W'}{v_t} \end{cases} \quad \text{(13)}$$

Where u_t is the total pixel length in the u-direction of the image plane and v_t is the total pixel length in the v-direction of the image plane. According to the equations (11), (12) and (13), the target coordinates on the image plane have the following correspondence with its coordinates of the physical coordinate system:

$$\begin{cases} x = 2L * \dfrac{f}{H - h_y} * \dfrac{(u_0 - u)}{u_t} \\ y = W * \dfrac{f}{H - h_y} * \dfrac{(v - v_0)}{v_t} \end{cases} \quad \text{(14)}$$

(x, y), which is calculated by equation (14), is the position of the target on the two-dimensional image under the information acquisition system plat for m of this paper.

5.2. The Three-Dimensional Coordinate Positioning Based on Two-Dimensional Imaging System

In order to realize the three-dimensional positioning based on two-dimensional imaging, this paper restores the coordinates of the target in the actual three-dimensional space according to the coordinates on the image plane obtained by the equation (14).

As shown in Figure **8**, $A(X_A, Y_A, Z_A)$ is the intersection of light and the water surface, $S'(X_{S'}, Y_{S'}, Z_{S'})$ is the coordinates of the real fish's virtual image, $S''(X_{S''}, Y_{S''}, Z_{S''})$ is the coordinates of the virtual fish's virtual image. On the image plane, the images of S' and S'' are denoted as $u(x_u, y_u)$ and $u'(x_{u'}, y_{u'})$ in the image physical coordinate system, respectively, and $u(u,v)$ and $u'(u',v')$ in the image pixel coordinate system, respectively. According to the similar triangular principle and plane coplanar equation, A , S', S'' , u and u' exist the following correspondence:

$$\begin{cases} \frac{H-h}{f} = \frac{X_A}{x_u} = \frac{Y_A}{y_u} \\ \frac{H-Z_{S'}}{f} = \frac{X_{S'}}{x_u} = \frac{Y_{S'}}{y_u} \\ \frac{X_{S''}}{Y_{S''}} = \frac{x_{u'}}{y_{u'}} \\ \frac{X_{S'}}{Y_{S'}} = \frac{x_u}{y_u} \end{cases} \tag{15}$$

Combining the equation (14) and the first and the second equation so f equation (15), the coordinates of A conform to the following relationship:

$$\begin{cases} X_A = 2L * \frac{H-h}{H-h_y} * \frac{u_0 - u}{u_t} \\ Y_A = W * \frac{H-h}{H-h_y} * \frac{v - v_0}{v_t} \end{cases} \tag{16}$$

And the coordinates of S' conform to the following relationship:

$$\begin{cases} X_{S'} = 2L * \frac{H-Z_{S'}}{H-h_y} * \frac{u_0 - u}{u_t} \\ Y_{S'} = W * \frac{H-Z_{S'}}{H-h_y} * \frac{v - v_0}{v_t} \end{cases} \tag{17}$$

Substituting equations (5) and equations (14) into the third and the fourth equations of equation (15), the following correspondence is obtained:

$$\begin{cases} \frac{Z_S}{Y_S} = \frac{2L * (u' - u_0) * v_t}{W * (v' - v_0) * u_t} \\ \frac{X_{S'}}{Y_{S'}} = \frac{2L * (u_0 - u) * v_t}{W * (v - v_0) * u_t} \end{cases} \tag{18}$$

Combining the equation (5) (6) (7) (8) (9) (16) (17) (18), the coordinate value of $S(X_S, Y_S, Z_S)$ can be calculated, which is the actual position of real fish in three-dimensional space.

5.3. The Specific Restoration Process of Three-Dimensional Coordinate Positioning Algorithm

Based on the information acquisition system in this paper, some data can be measured directly, such as the focal length of the camera is f (cm), the height of the camera from the tank bottom is H (cm), the tank's length, width and height are L (cm), W (cm)and h (cm), the height of the water surface is h_y (cm). And the pixel length and width of the captured video frame image are easily obtained. Using these known quantities to obtain the three-dimensional coordinates of the fish target is the ultimate goal of this experiment. Specific algorithm steps are as follows:

Step 1: Initialize the data. Such as the camera height H (cm), the water tank length L (cm), width W (cm), height h (cm), the height of the water surface h_y (cm) and the pixel length and width of the video frame image.

Step 2: Use image segmentation technology to segment the target and accurately locate and extract its center point which represents the coordinates of the target in the image plane.

Step 3: Calculate the coordinates of the target point in the image physical coordinate system based on the two-dimensional imaging process.

Step 4: According to the correspondence relationship between the two-dimensional coordinates and the three-dimensional coordinates, establish the correlation equations of the points' coordinates.

Step 5: According to the above correlation equation, calculate the solution. This solution is the three-dimensional coordinates of the target. The end.

6. Simulation Experiments

6.1. Experimental Results

In this paper, the experimental tank's length, width and height are all 25cm, water depth is 20cm high, the height of camera from the bottom of the tank is 75cm. The experimental fish is red crucian carp. The water in the tank comes from tap water. In order to verify the accuracy of the three-dimensional coordinate positioning method of fish target, the two groups of experimental data were taken to simulate the experiment. One group used a swimming live fish, as shown in Figure **9**. The other group used a static artificial fish, as shown in Figure **10**.

Figure **9** shows five images of a 30-minute video, with numbers from left to right being **(a)**, **(b)**, **(c)**, **(d)**, and **(e)**. Using the three-dimensional coordinate positioning algorithm in this paper, the three-dimensional coordinates of the live fish in these images have been obtained. The results are shown in Table **1**. It can be seen from the fish positions in Figure **9** and the results in Table 1 that the calculated three-dimensional coordinates of the fish are consistent with their actual position. It is shown that the proposed method can effectively locate the three-dimensional coordinates of the target.

(a) (b) (c) (d) (e)

Fig. 9. Five images of live fish.

Table 1. 3D coordinates of live fish in Fig. 9 calculated by this method.

Images	**Three-dimensional coordinates of live fish**		
	X(cm)	**Y(cm)**	**Z(cm)**
Image(a)	0.71	-7.21	2.45
Image(b)	1.45	-2.30	2.09
Image(c)	1.13	2.58	2.34
Image(d)	21.63	-7.11	1.76
Image(e)	24.96	9.83	5.87

As the living fish has been swimming, it is difficult to measure the three-dimensional coordinates of live fish accurately in real time. Therefore, in order to further verify the accuracy of the algorithm proposed in this paper, we have done an artificial fish experiment. Since the artificial fish can be fixed at a certain position, the specific coordinates of the fish can be easily measured and compared with the coordinates calculated by the algorithm. In the measurement process, the highest point of the artificial fish is taken as the measurement point.

(a) (b) (c) (d) (e)

Fig. 10. Five images of artificial fish.

The artificial fish images are shown in Figure **10**, **(a)** shows that the artificial fish is in the upper right corner of the tank bottom, **(b)** shows that the artificial fish is in the middle of the right side of the tank bottom, **(c)** shows that the artificial fish is in the lower right corner of the tank bottom, **(d)** shows that the artificial fish is in the upper left corner of the tank bottom and **(e)** shows that the artificial fish in the lower left corner of the tank bottom. Table **2** shows the three-dimensional coordinates of the artificial fish calculated by our method and the coordinates we measured. It can be seen from Table **2** that the actual position of the artificial fish coincides with the three-dimensional coordinates calculated by the method, which proves that the proposed method of three-dimensional coordinate positioning method is correct.

Table 2. 3D coordinates of the artificial fish in Fig. 10: The coordinates calculated by this method and the coordinates of the measurements.

Image	the coordinates calculated by this method			the coordinates of the measurements		
	X(cm)	Y(cm)	Z(cm)	X(cm)	Y(cm)	Z(cm)
Image (a)	0.61	-9.30	1.77	0.58	-9.34	1.75
Image (b)	0.31	1.09	2.03	0.34	1.05	2.05
Image (c)	0.01	9.37	1.90	0.02	9.45	2.02
Image (d)	25.00	-9.30	1.68	25.03	-9.42	1.58
Image (e)	24.08	8.92	1.75	23.95	9.05	1.75

6.2. Performance Analysis and Comparison

In this paper, the experiment is in the CPU running speed of 3.4GHz, memory capacity of 1.88GB, and Matlab R2009a software environment. The average time to run a video image is 0.00508735 seconds. If the three-dimensional positioning software is developed by Visual C++, its speed will be further improved. Therefore, this tracking and positioning method can guarantee the real-time of water quality detection and early warning.

Table 3. The comparisons of this method with the early methods.

Methods	Number of cameras	Number of mirrors	Account external factors?	Data storage form
Visdcido[4]	2	0	no	2 groups of 2D date
Wu[6]	2	1	no	2 groups of 2D date
Pereira[7]	1	1	no	3D data
Xu[9]	1	1	no	3D data
Zhu[10]	1	2	no	3D data
This method	1	1	yes	3D data

Table **3** shows the positioning method proposed in this paper and the data collection method of the water quality detection system proposed by the early literature. It can be seen from Table 3 that the positioning method proposed in this paper has the following advantages:

1. Compared with Visdcido's [4] and Wu's [6] work, this positioning method used a single camera while they used a dual camera. In the same time, and the camera

acquisition rate of the same circumstances, this method requires a small amount of data to dcal with, which can save the computer's memory consumption.

2. This paper fully considered the effect of water refractive index on the three-dimensional positioning accuracy, and Pereira [7] did not consider this problem.

3. Xu [9] gave a single camera single plane mirror positioning method, and the plane mirror was placed on the upper side of the tank. As the water surface was often lower than the height of the tank, the collected images tend to have a blank intermediate band, reducing the target resolution. In this paper, the camera was placed above the tank and the plane mirror was placed in the side of the tank, which make up for their lack.

4. Zhu [10] proposed two plane mirrors and a single camera to construct a three-dimensional target analysis system. They placed two plane mirrors on the top and left sides of the tank respectively, and the camera was placed in front of the tank. The images were divided into four parts, on the upper left was the blank part, on the upper right was the top plane mirror, on the lower left was the other plane mirror, on the lower right was the fish. In this way, the three-dimensional coordinate of the target was synthesized to describe the spatial trajectory of fish. As the quarter of the image is blank, the resolution of the target is reduced, and the loss of target behavior information is likely to occur.

7. Conclusion

This paper presents a method to realize the three-dimensional coordinate positioning of fish target based on single camera. A single camera and a single plane mirror were used to build the acquisition equipment. According to the principle of underwater object imaging, the principle of plane mirror imaging, the principle of camera imaging and image processing technology, this three-dimensional coordinate positioning method was deduced theoretically. Then the method was validated by live fish and artificial fish experiments. The results showed that the method was accurate and effective.

The main factors that affect the accuracy of this method are: camera position, image segmentation and target point extraction. The camera's position can be calibrated with high precision, so as to avoid the error caused by imaging. Image segmentation technology is the key to extract the target, according to the image segmentation technology used in this paper can reasonably extract the effective target. The extraction of the target point means extracting one point to represent the whole object, which is often selected by the center or the gravity center. In this paper, the experimental object is small and need to be corroded after determining the target, so extracting the center point can effectively reduce the error to improve the accuracy. But how to extract more effective target point need further study.

The three-dimensional coordinate positioning method proposed in this paper can restore the three-dimensional coordinate information of live fish in real time. Based on the acquired three-dimensional coordinate information, we can calculate the trajectory, velocity and acceleration of the fish target in real time, and provide a reliable basis for further study on the behavior characteristics of the fish target, which lays the foundation for the study of the behavioral characteristics of the fish target in the three-dimensional space.

Acknowledgements

This work is supported by the National Natural Science Foundation of China (Nos. 61272310, 61573316), the ZheJiang province Natural Science Foundation of China (Nos. LY15F020032, LQ15E050006).

References

1. Vowles AS, Anderson JJ. (2014). Effects of avoidance behaviour on downstream fish passage through areas of accelerating flow when light and dark. Animal Behaviour, 92, 101-109. doi:10.1016/j.anbehav.2014.03.006.
2. Weiqing M, Zuwei W, Beibei H. (2016). Heavy metals in soil and plants after long-term sewage irrigation at Tianjin China: A case study assessment. Agricultural Water Management, 171, 153-162. doi:10.1016/j.agwat.2016.03.013.
3. Lu S, Wang J, Pei L. (2016). Study on the effects of irrigation with reclaimed water on the content and distribution of heavy metals in soil. International Journal of Environmental Research and Public Health, 13(3), 298-305. doi:10.3390/ijerph13030298
4. Visdcido SV, Parris JK, Grunbaum D. (2004). Individual behavior and emergent properties of fish schools: a comparison of observation and theory. Marine Ecology Progress Series, 273, 239-249. doi:10.3354/meps273239.
5. Nimkerdphol K, Nakagawa M. (2008). Effect of sodium hypochlorite on zebrafish swimming behavior estimated by fractal dimension analysis. Journal of Bioscience and Bioengineering, 105(5), 486-492. doi:10.1263/jbb.105.486.
6. Wu GH, Zeng LJ. (2007). Video tracking method for three-dimensional measurement of a free-swimming fish. Science in China Series G: Physics, Mechanics and Astronomy, 50(6), 779-786. doi:10.1007/s11433-007-0071-5.
7. Pereira P, Oliveira RF. (1994). A simple method using a single video camera to determine the three-dimensional position of a fish. Behavior Research Methods, Instruments, & Computers, 26(4), 443-446.
8. Derry JF, Elliott CJH. (1997). Automated 3-D tracking of video-captured movement using the example of an aquatic mollusk. Behavior Research Methods. Instruments. & Computers, 29(3), 353-357.
9. Xu P, Han J, Tong J. (2012). Preliminary studies on an automated 3D fish tracking method based on a single video camera (in Chinese). Journal of Fisheries of China, 36(4), 623-628. doi:10.3724/SP.J.1231.2012.27439.
10. Zhu L, Weng W. (2007). Catadioptric stereo-vision system for the real-time monitoring of 3D behavior in aquatic animals. Physiology & Behavior, 91(1), 106-119. doi:10.1016/j.physbeh.2007.01.023.
11. Xia Q, Que S. (1999). Visual depth underwater objects (in Chinese). Physics Bulletin, 01, 13-14.
12. Mu G, Zhan Y. (1978). Optics (in Chinese). People's Education Press. BeiJing.

13. Yuan G, Xue M, Han Y, Zhou P. (2010). Mean shift object tracking based on adaptive multi-features fusion (in Chinese). Journal of Computer Research and Development, 47(9), 1663-1671.
14. Luo H, Du F, Kong F. (2015). Pixel feature-weighted scale-adaptive object tracking algorithm (in Chinese). Journal on Communications, 36(10), 200-2010. doi:10.11959/j.issn.1000-436x.2015259.
15. Fan C, Ouyang Ho, Zhang Y. (2013). Small probability strategy based otsu thresholding method for image segmentation (in Chinese). Journal of Electronics & Information Technology, 35(9), 2081-2087. Doi:10.3724/SP.J.1146.2012.0159.
16. Liu X, Lu Y, Lei L, Qu Z. (2015). A multi-objective image segmentation method based on fcm and discrete regularization (in Chinese). Journal of Computer-Aided Design & Computer Graphics, 27(1), 142-146.
17. Siang Tan K, Mat Isa NA. (2011). Color image segmentation using histogram thresholding–Fuzzy C-means hybrid approach. Pattern Recognition, 44(1), 1-15. doi:10.1016/j.patcog.2010.07.013.
18. Zhang Z, Shui P. (2015). SAR image segmentation algorithm using hierarchical region merging with edge penalty (in Chinese). Journal of Electronics & Information Technology, 37(2), 261-267. doi 10.11999/JEIT14033l.
19. Wang J, Xu J. (2009). Camera calibration method based on planar pattern (in Chinese). Computer Engineering and Design, (1), 259-260.
20. Song T, Chu G-Y, Hou P-G, Li H-B, Chen C. (2016). Fisheye camera calibration algorithm optimized by centroid method (in Chinese). Acta Photonica Sinica, 45(5), 1-6. Doi:10.3788/gzxb20164505.0512005.
21. Lu G, Chen H, and Xiao L, Su H, Zhong R. (2016). Multi-view video stitching in surround view parking assist system (in Chinese). Journal of Nanjing University of Posts and Telecommunications(Natural Science Edition), 36(3), 10-17. doi:10.14132/j.cnki.1673-5439.2016.03.002.
22. Herrera C, Kannala J, Heikkilä J. (2012). Joint depth and color camera calibration with distortion correction. IEEE Transactions on Pattern Analysis and Machine Intelligence, 34(10), 2058-2064. doi:10.1109/TPAMI.2012.125
23. Tian M. (2006). A survey of calibration in a vision-robot system (in Chinese). Industrial Instrumentation & Automation, (2), 14-17.

Discrete Event Models in Telemedicine

Calin Ciufudean

"Stefan cel Mare" University, 9 Universitatii str.,
720225, Suceava, Romania
calin@eed.usv.ro

Abstract. Communication systems are made of reliable components, both hardware and software, and therefore they may have only a few failures in exploitation. For an application field such as telemedicine this desiderate is a must. We model these systems dedicated to telemedicine with randomized pulse modulation data traffic using a finite Markov chains. Based on previous assumption telemedicine communication systems failures are considered, and therefore modelled using the rare event with a finite-state formalism. Our work is focused on improving the telemedicine systems` reliability using randomized pulse width modulation.

Keywords: Telemedicine, Markov chains, pulse width modulation, rare events, discrete event systems.

1. Introduction

In order to estimate patterns of telemedicine communication systems we deal with discrete event models of their reliability based on observed success/failure data as we noticed in practice. As noticed in literature, failures are seen as a rare events, in telemedicine communications [1-3], and therefore we assume that their occurrence probabilities are smaller by at least several orders of magnitude than probabilities of ordinary failure events. Therefore, the mean time to failure (MTTF) one of the most used reliability parameter, is a large number. User friendly representation and computations characterise the Markov chain for modelling and analysing rare events. The Markov chain provides also a direct measure of rarity using steady-states probabilities. Here we deal with a discrete-parameter, finite-state Markov chain [4, 5] used for representing both data communications failure (as transitions to a rare fail-state) and randomized pulse width modulation schemes of transmitter. We choose this approach as it offers few advantages, such as: the randomized modulation of switching reduces filtering equipment's and allows an explicit time domain control, and also is reported in literature, randomized modulation is effective for dealing with narrow band constraints [6]. The remainder of this paper is as follows: Section 2 describes the basic principle of the proposed data transmitter and receiver, Section 3 deals with our approach for Markov chains modelling formalisms of randomized modulation availability. Section 4 exemplifies a randomized modulation modelled with periodic Markov chains, and proposes a new model of periodic Markov chain. Section 5, based on an illustrative example, analysis the reliability of this approach here entitled reliability Markov chain model (RMCM). Conclusion synthesizes the paper, the proposed model and display few development slopes.

2. Proposed Data Structure

Basically, a data transmitter involves generating a switching function f(t), which for example is equal to "1" when the switch is conducting and the "0" otherwise. This is schematically represented in Figure **1**. We assume that for value "1" there is some medical data to be transmitted; respectively there is one package (which includes several distinct diagnosis, images, medication *etc*., each one embedded in a sine wave) of data to be transmitted, as illustrated in Figure **2**. For verifying the correctness of the received data, the decoder counts the number of sine waves periods which fits the length of the pause. The resulted number, which must be a par multiple of the sinusoids (*e.g*., a par multiple is needed in order to verify if one sinusoid is missing due to perturbations), indicates the specific number and the sort of data inside the transmitted package, such as "n" medical analyses for patient John Doe. As shown in Figure **2** the number of the codified pulses g(t) determines the switching function f(t) of the data transmitter [4]. Therefore the data security is double checked and protected against electromagnetic interferences, as well against data loss or attenuation throughout passing reactive circuits.

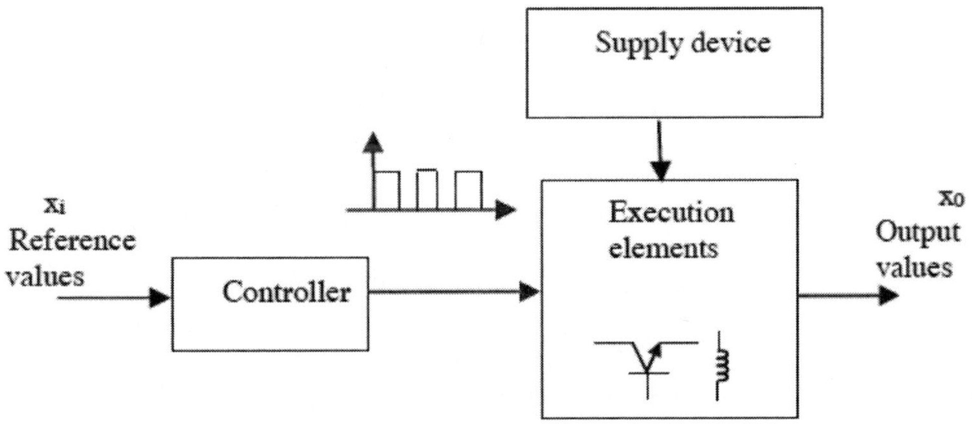

Fig. 1. Schematic representation of the proposed switching transmitter.

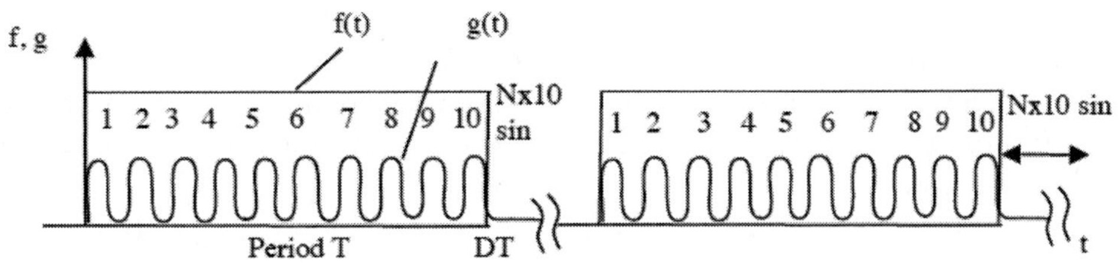

Fig. 2. Switching functions of packages of data traffic in telemedicine.

It is stated in literature [4-8] that the average value or duty ratio D for f(t) usually determines the nominal output of a transmitter, as wave forms that are periodic have spectral components only at integer multiples of the fundamental frequency. We noticed that an effort was directed toward the optimization of deterministic PWM waveforms, and alternative in the form of randomized modulation for d.c./a.c. and a.c./a.c. conversion is based on schemes in which successive randomizations of the periodic segments of the switching pulse train are statistically independent and governed by probabilistic rules. These schemes are denoted in literature as stationary [3]. Next chapter describes an approach for the synthesis of this class of stationary randomized modulation schemes that enables explicit control of the time-domain performance of the data traffic applied in telemedicine.

3. Estimating Switching Availability with Markov Chains

Telemedicine data transmission process deals with the switching code system availability in order to design a randomized switching procedure that minimizes given criteria for spectral characteristic of f(t). For estimating the data transmission system availability we introduce a failure state due to imperfect data coverage (denoted here by "c") and repair (denoted here by "r") [4]. In order to explain the impact of data imperfect coverage we consider the coding system in Figure **3** which includes two identical switching codes device (SWCD).

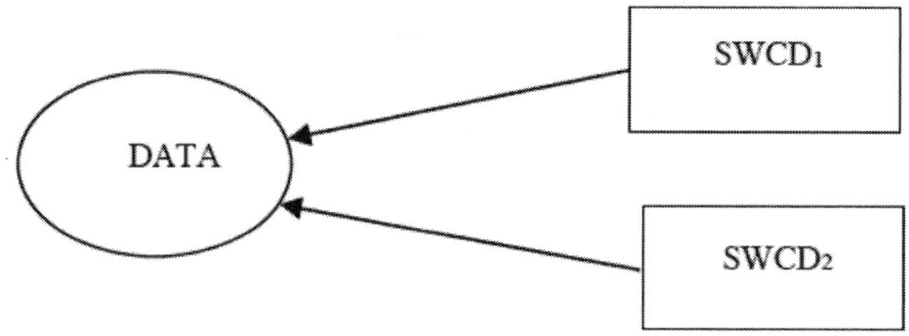

Fig. 3. Example of perfect coverage data with two identical codes.

For exemplifying the coverage of a transmission data system we notice that if the coverage of the system is sufficient, *e.g.*, c = 1, then data is transmitted correctly as long as one of the $SWCD_i$ is operational, where i = 1, 2. If the coverage is insufficient, then data is incorrect with probability 1 – c (*i.e.*, we may say that data fail with probability 1 – c).

The Markov chain for modelling data using $SWCD_i$, where i = 1, …, k, with k being a nonnegative integer is shown in Figure **4**, where parameters λ, μ, c, r denote respectively the failure code rate, repair code rate, coverage data rate, and the successful data repair rate of SWCD.

The first part of the horizontal transition rate with term 1 – c represents the data (*e.g.*, code) failure due to imperfect coverage of an alternative SWCD. The second part, with the term 1 – r represents imprecise repair of SWCD.

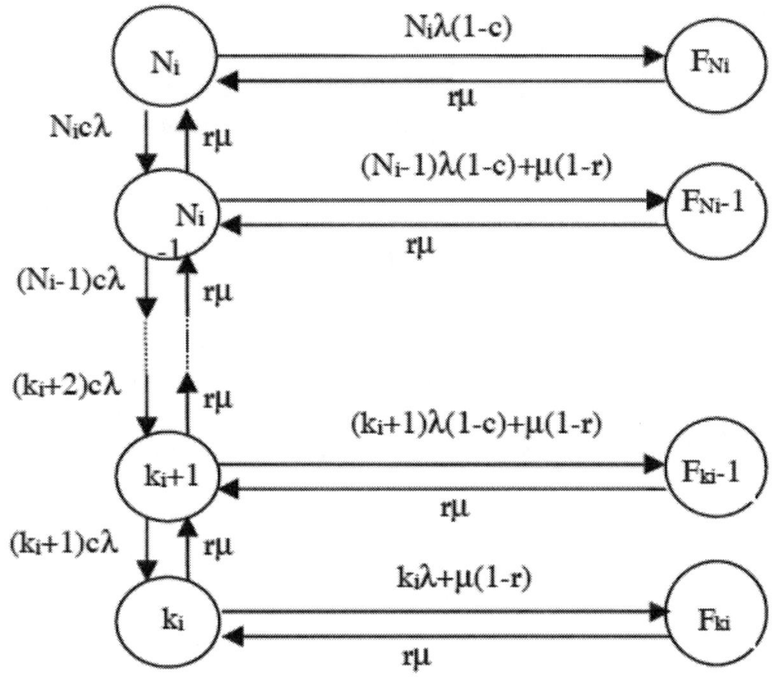

Fig. 4. Markov model for $SWCD_i$.

The down/up arcs model the failure/repair of $SWCD_i$. As we deal with rare events, we say that only one code fails at a time in a $SWCD_i$. In Figure **4** on state N_i the $SWCD_i$ is functioning with all N_i codes operational. The trajectory of our system, respectively the states of $SWCD_i$ changes from one of the k_i working states to one of the F_{ki} codes failure states due to imperfect coverage (1 - c), or due to imperfect repair (1 - r). We notice that perfect fault/repair coverage of the system reduces the Markov chain to one-dimension model. The solution of the Markov chain model given in Figure 4, and implicitly the solution for system functionality is given by the probability that at least k_i codes are correct at time t. The availability of $SWCD_i$ is determined using the following relation [13]:

$$A_i(t) = \sum_{k_i}^{N_i} P_{k_i}(t), \quad \text{for } i = 1, 2, \ldots, n \tag{1}$$

where, $A_i(t)$ = the availability of $SWCD_i$ at time t; N_i = the total number of codes in $SWCD_i$; k_i = required minimum number of operational codes in $SWCD_i$.

After a Markov chain for $SWCD_i$ is built and desired probabilities $A_i(t)$, i = 1, 2, … , n are determined, the system`s availability (*e.g.*, the switching code system) is given by the following relation:

$$A(t) = \max A_i(t) A(t) = \max A_i(t) \tag{2}$$

In order to optimize the data transmission is mandatory to minimize the discrete spectral components (denoted as narrow-band optimization), and to ensure minimization

of signal power in a given frequency range (denoted as wide-band optimization) [14]. We assume that our Markov model goes through a sequence of n classes of states C_i , occupying a state in each class for an average time σ_i, i = 1, ..., n.
The time-average autocorrelation of the random process f(t) is defined as [6]:

$$R_f(\tau) = \lim_{n \to \infty} \frac{1}{2w} \int_{-w}^{w} E\,[f(t) \cdot f(\tau + t)] \cdot dt \qquad (3)$$

where the expectation E[.] refers to the whole ensemble [.].

The Markov chain of class C_k, with the time-averaged autocorrelation (3) is scaled by $\tau_k / \sum_{i=1}^{n} \tau_i$, where τ_i the expected time is spent in the class C_i before a transition into the class C_{i+1}. We define the (n x n) state-transition matrix P with (k, i)th entry given by the probability that at the next transition the chain goes to state i, given that it is currently in state k. P is a stochastic matrix, so it has a single eigenvalue $\lambda_i = 1$, with corresponding eigenvector $1_n = [1\ 1\ \dots\ 1]$, and all other eigenvalues less than 1. In [1] it is proven that after a possible remembering of the states, the matrix P has a block-cyclic form:

$$P = \begin{bmatrix} 0 & P_{12} & 0 & \dots & 0 \\ 0 & 0 & P_{23} & \dots & 0 \\ \dots & \dots & \dots & \dots & \dots \\ 0 & 0 & 0 & \dots & P_{n-1,n} \\ P_{n1} & 0 & 0 & \dots & 0 \end{bmatrix} \qquad (4)$$

Let P_P denote the product of submatrices of P: $P_P = P_{n1} \dots P_{23} \,.\, P_{12}$, and let v_i denote the vector of steady-state probabilities, conditional on the system being in class C_i. We have:

$$V_i^* = v_i \cdot P_P \qquad (5)$$

The time spent in class C_i (maintenance time) is:

$$\tau_i = \sum_k V_k^* \cdot \tau_k \qquad (6)$$

Let $T_o = \sum_{i=1}^{n} \tau_i$ and let Θ_i = diag (V_i*). If the first data string belongs to the class i, then the pulse $\tau_i + \tau$ belongs to the class (i + m)/n, where m represents the number of transitions between data strings between moments τ_i and $\tau_i + \tau$. We have a particular case with average duration of some classes' null, which means that these classes are avoided [13-15]. This construction deals with a simplified Markov chain model (we present this assumption, which we believe to be novel, in the next section). When we add the contribution of all classes to the average power spectrum (scaled by the relative average duration of each class) the result can be written as follows [4]:

$$S(f)=\frac{1}{T_0}\left[\sum_{i=1}^{n}\frac{\tau_i}{T_0}\cdot U_i^T(f)\cdot\Theta_i U_i(f)+2R_e\left(1_n^T\cdot S_C\cdot 1_n\right)\right]+\frac{1}{T_0^2}R_e\left(1_n^T\cdot S_d\cdot 1_n\right)\sum_{i=-\infty}^{\infty}\delta\left(f-\frac{i}{T^*}\right) \tag{7}$$

Where T^* is the greatest common divisor of all waveform duration, 1_n is an n x 1 vector of 1 and U_i is the vector of Fourier transforms of waveforms assigned to states in class C_i. The matrix $S_{c,i}$ has a Toeplitz structure, with (k, i)th entry:

$$S_{ck,i}(f)=\frac{\tau_i}{T_0}\cdot U_k{}^T(f)\cdot(I-\Lambda_k(f))^{-1}\cdot\Lambda_{k,i}(f)\cdot U_i(f) \tag{8}$$

where Λ_k is a product of n matrices: $\Lambda_k=Q_{k-1,k}, \ldots, Q_{k,k+1}$, and $\Lambda_{k,j}=Q_{k,k+1}, \ldots, Q_{i-1,i}$. Where Q is a matrix n x n whose (k, i) entry is $Q_{k,i}(\sigma) = P_{k,i}\delta(\sigma-\tau_k)$. Also, the (k, i)th entry of S_d is given by relation:

$$S_{dk,i}(f)=\frac{\tau_i}{T_0}\cdot U_k^T(f)\cdot V_k\cdot(V_i)^T\cdot U_i(f) \tag{9}$$

Starting from the notion of truncated Markov chains with absorbing states [6], we propose a simplified model of Markov chains for random modulation. The proposed Markov chain $X^{(m)}$ has the states {m, m+1, … } aggregated into an absorbing class of states, with the transition matrix ${}_{(m)}T$:

$${}_{(m)}T=\begin{bmatrix} {}_{(m)}P & p_m \\ 0 & 1 \end{bmatrix} \tag{10}$$

Where $p_m=\left[\sum_{j\geq m}p_{1j},\sum_{j\geq m}p_{2j},\ldots,\sum_{j\geq m}p_{m-1,j}\right]$.

Since ${}_{(m)}P$ is irreducible for all m, it follows that $X^{(m)}$ constitutes an irreducible Markov chain for all m, where m is an absorbing class of states [m]. The n-step transition matrix ${}_{(m)}T^n$ is:

$${}_{(m)}T^n=\begin{bmatrix} {}_{(m)}P^n & \left(I_m+{}_{(m)}P+{}_{(m)}P^2+\ldots+{}_{(m)}P^{m-1}\right)p_m \\ 0 & 1 \end{bmatrix} \tag{11}$$

Where ${}_{(m)}P^n = {}_{(m)}p_{ij}{}^n = p_{ij}(n)$.

The transition probability between realizable classes of states is 1. The transition matrix probability also indicates the priorities between the states of the system.

4. Illustrative Example

In this section we deal with an illustrative model applicable for the data transmitter`s code depicted in Figure **2** in order to generate a switching function where blocks of pulses have deterministic duty ratios: [0.75, 0.5, 0.25]. The periodic Markov chain shown in Figure **3**, with six states divided into three classes, is an example of a solution to this issue [6, 7]. A short cycle (duration 3/4) and a long cycle (duration 5/4) are available in each of the four classes. According to the theoretical approach in the previous paragraph, we build a simplified Markov chain in Figure **4**. The Markov chains in Figure **5**, respectively in Figure **6** have the same transition probabilities between states S_i, i=1,…, 6 of classes C_j, j=1,…, 3. One can observe that the Markov chain in Figure **6** is more intuitive and tidy than the one in Figure **5**. The transition probabilities equal to 1 in Figure **6** are conditioned by the existence of transition between the states of different classes.

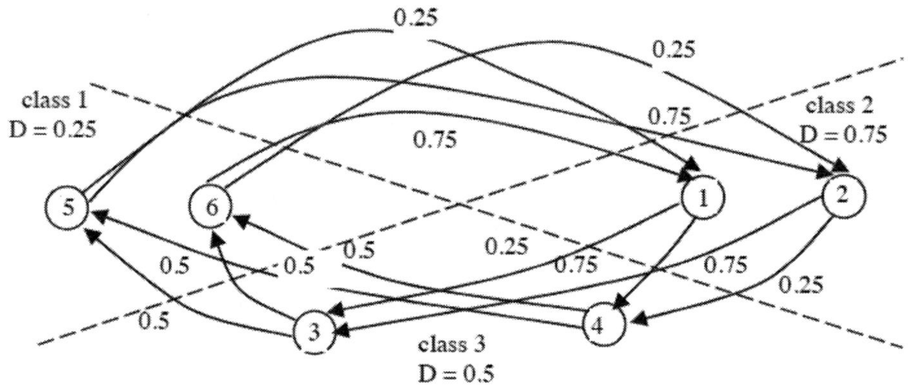

Fig. 5. Classic Markov chain for modeling the switching example.

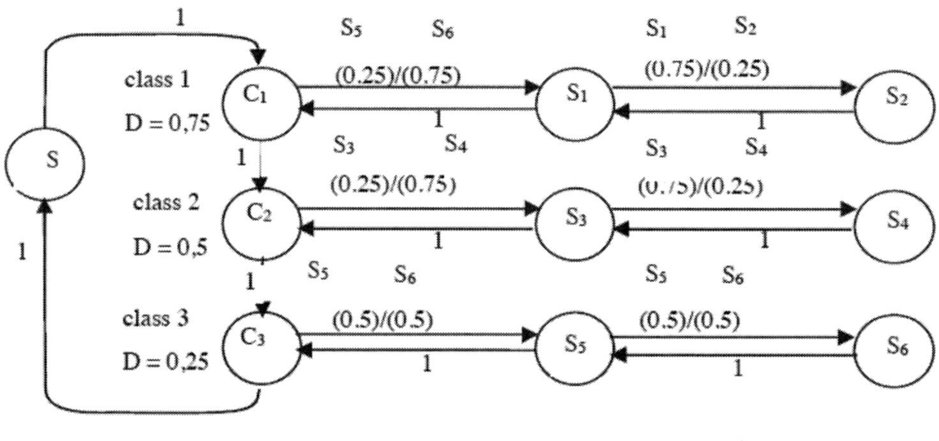

Fig. 6. Proposed Markov chain for modeling the switching example.

We analyze this Markov chain with the help of equation (4) and we compare the theoretical predictions with the estimates obtained in Monte Carlo simulations. The agreement between the two is quite satisfactory: the theoretical prediction for the impulse strength at f = 4 is 0.0035, and the estimated value is 0.0037. The Markov chain in Figure **4** allows dealing with many more classes as graphical representation is simplified and can significantly improve the tractability of the optimization of the multi-state Markov chains.

6. Reliability Estimation of our Model

The Markov chain for estimating the reliability of our model has at least three states: starting-state S, working state W, and fail state F. State sequences are realizations of reliability Markov chain model (RMCM) [7, 8, 15]. A realization from S to first occurrence of W represents a single successful execution cycle of data transmission. A transition from any state to F represents a failure of data transmission. The probabilities on arcs in RMCM are the values estimated for the usage profile and component reliabilities expected in practice [9]. We notice that if state i RMCM, and $i \neq F$, has been visited n_i times and exited without failure, then the probability of failure at state i is no greater than $1/(n_i + 1)$. Let random variable n_F be the number of visits to F in a randomly generated realization of n transitions starting in a state S. Let $\lambda=E(n_F)$ be the mean value of the probability low of n_F. Let $P = [p_{ij}]$ denote the RMCM's transition probability matrix. The RMCM have all states reachable from S by traces with nonzero probability and arcs from both F and W to S with $P_{FS} = P_{WS} = 1$, so that a successful or unsuccessful path terminated in W, respectively in F state causes an immediate restart in initial state S. RMCM's steady-state probability distribution $\Pi = [\pi_S, \ldots, \pi_F]$ is the unique solution of $\Pi = \Pi P$ where $\Sigma\, \Pi_i = 1$ and $\pi_i > 0$ is the limiting relative frequency of occurrence of state i as a count transition, *e.g.*, recurrence time [4, 16, 17]. Adopting a Poisson law with parameter λ, developments in small number laws stress that $P_0(\lambda)$ is an approximation and compute an upper bound for measuring the distance between $P_0(\lambda)$ and the probability law $L(n_F)$. The total variation distance $d_{TV}\,[L(n_F), P_0(\lambda)]$ is defined as show in relation (12) for events A in the sample space [10], where $P_0(\lambda)$ is the Poisson distribution with parameter λ:

$$d_{TV}\left[L(n_F),\, P_0(\lambda)\right] = \sup_A \left|L(n_F)(A) - P_0(\lambda)(A)\right| \tag{12}$$

Since Π_F equals the limiting relative frequency of state F, for large n we have:

$$\Pi_{F} \approx \frac{E(n_F)}{n} = \frac{\lambda}{n} \tag{13}$$

We may approximate for large n, and rare state F:

$$L(n_F) \approx \frac{(n\Pi_F)^k}{k!} \cdot e^{-n\Pi_F} \tag{14}$$

We observe that $\lambda \approx n \cdot \Pi_F$ is the approximate parameter for a full sequence of transition, not per transition. Since the mean sequence length is m_{SS} transitions, the expected count of transitions between visits to F is [4, 10, 18]:

$$m_{SS} E(\lambda) = m_{SS} \frac{\pi_S}{\pi_F} = m_{SS} \frac{m_{FF}}{m_{SS}} = m_{FF} \tag{15}$$

We exemplify this approach on the three state Markov chain (RMCM) given in Figure **5**. This RMCM corresponds to the Markov chain given in Figure **4**, where we added the fail state F associated to the states C_i, $i = 1, 2, 3$ associated to the classes in the Markov chain model for pulse width modulation scheme discussed in section 3.

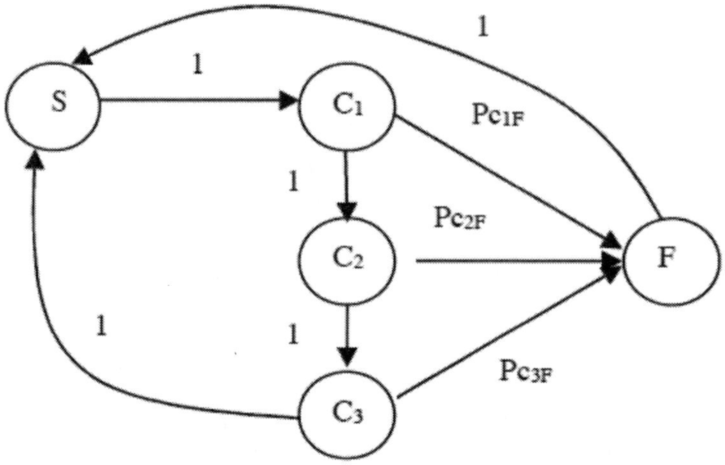

Fig. 7. Five state RMCM associated to the Markov Chain in Fig. **6**.

By colligating the RMCM in Figure **5** with the Markov chain in Figure **6**, we notice the transition probabilities for the ordinary usage-states given in Figure **6** (p_{ij}) and the small probabilities Pc_{iF}, i = 1,2,3 in Figure **7**. Given that for highly reliable security communication systems we may have $0 \leq Pc_{iF} \leq 10^{-3}$, and correspondingly $0 < \Pi_{Fi} \leq 1{,}3 \cdot 10^{-4}$, which allow us to presume that for $Pc_{iF} \approx 10^{-4}$ the stationary distribution vector:

$$\Pi = [\pi_S, \pi_{C1}, \pi_{C2}, \pi_{C3}, \pi_F] = [0.14278,\ 0.42813,\ 0.14283,\ 0.14276,\ 1.42689 \cdot 10^{-5}] \quad (16)$$

The vector of mean recurrence time is:

$$[m_{SS}, ..., m_{FF}] = \left[\frac{1}{\pi_S}, ..., \frac{1}{\pi_F}\right] = [7.00400,\ 2.33412,\ 7.00120,\ 7.00470,\ 7.00821 \cdot 10^4] \quad (17)$$

The vector of the expected number of occurrences of states between transitions to non-rare state S is:

$$\frac{\pi}{\pi_S} = [1,\ 3.0009,\ 1.0005,\ 1.0005,\ 0.9989,\ 0{,}0005] \quad (18)$$

The vector of the expected number of occurrences of states between transitions to rare state F is:

$$\frac{\pi}{\pi_F} = [1.0009,\ 3.0047,\ 1.0009,\ 1.0012,\ 0.0003] \cdot 10^4 \quad (19)$$

The mean recurrence time of state S is $m_{SS} \approx 9$; therefore $n = 9 \cdot 10^4$ transitions correspond to approximatively 10^4 average sequences from S back to S (*e.g.*, we deal with reversible processes). The Poisson approximation of $L(n_F)$, see relation (11) for $n = 9 \cdot 10^4$ and $\lambda = 9\pi_F \cdot 10^4$ stands $p_{CF} = 10^{-4}$ and the rare state F has steady-state probability

$\pi_F \approx 1.43 \cdot 10^{-5}$, and the upper bound $0,903 \cdot 10^{-5}$, where $k = 0, 1, 2, 3$. The MTTF is $m_{FF} \approx 9 \cdot 10^4$ for $p_{CF} = 10^{-4}$.

7. Conclusion

Our paper focuses on synthesizing results of randomized modulation with Markov chains, suitable for data transmitter, for example the ones used in telemedicine. Randomized modulation switching schemes governed by Markov chains applicable to d.c./a.c. or d.c./d.c. converters have been described.

Our representation for complex periodic Markov chains we believed to be novel. We also believe that this new model offers a new perspective for the spectral characteristics and other associated waveforms in a converter to the probabilistic structure that governs the dithering of an underlying deterministic nominal switching pattern.

Further research will continue to focus on minimization of one or multiple discrete harmonics. This approach corresponds to cases where the narrow-band characteristics corresponding to discrete harmonics are harmful, as for example in the telemedicine traffic security. We also discussed results in rare events for security communication system based on finite-state, discrete-parameter, recurrent Markov chain, here entitled reliability Markov chain model (RMCM).

The chain provides a simple definition of failure as a rare event, respectively as a failure state F for which the steady-state Π_F is orders of magnitude smaller than Π_K for $k \neq F$, usually states of RMCM. Poisson law distribution bounds the transitions to a rare-fail state F in arbitrarily large size RMCM.

Further research will focus on improvement of the analytic capabilities of RMCM in the study of extreme values of rare events [19, 20] when failure is infrequent and mean time to failure (MTTF) is long.

References

1. Aldous D. (1999). Probability approximations *via* the poisson champing heuristics, Springer-Verlang, NY.
2. Thomson MG, Wittaker JA. (1996). Rare failure state in a markov chain model for software reliability. IEEE Trans Reliab, 48(2), pp. 107-115.
3. Evans GW, Gor TB, Unger E. (1996). A simulation model for evaluating personnel schedules in a hospital emergency department. WSC '96 Proceedings of the 28th conference on Winter simulation, pp. 1205-1209, IEEE Computer Society Washington, DC, USA.
4. Balloni AJ. (2004). Why GESITI? - Why management of system and information technology? Book Series IFIP International Federation for Information Processing, Publisher Springer/Boston, Book Virtual Enterprises and Collaborative Networks.
5. Mininisterio da Ciencia, Tecnologia e Inovacao Brasil. (2011). An evaluation of the management information system and technology in hospitals.
6. Carter MW, Lapierre SD. (2001). Scheduling emergency room physicians.

Health Care Management Science 4, 347-360, Kluwer Academic Publishers.
7. Burke EK, De Causmaecker P, Berghe GV, Van Landeghem H. (2004). The state of the art of nurse rostering. Journal of Scheduling 7, pp. 441- 499, Kluwer Academic Publishers.
8. Lo CC, Lin TH. (2011). A particle swarm optimization approach for physician scheduling in a Hospital Emergency Department. Seventh International Conference on Natural Computation (ICNC), vol. 4, pp. 1929-1933.
9. Daknou A, Zgaya H, Hammadi S, Hubert H. (2010). A dynamic patient scheduling at the emergency department in hospitals, IEEE Workshop on Health Care Management (WHCM), Vol. 1, pp. 1-6.
10. Moses Y, Tennenholtz M. (2002). Artificial Social Systems, www.Kluivert.dl., Sistem Automato.
11. Zurawski R, Zhon M. (1994). Petri nets and industrial applications; A tutorial. IEEE Trans Ind Electron, vol.41, 567-583.
12. Zurawski R. (1994). Systematic construction of functional abstraction of Petri nets model of flexible manufacturing systems. IEEE Trans Ind Electron, Vol.41, 584-592.
13. Ciufudean C, Petrescu C, Filote C. (2005). Performance evaluation of distributed systems. 5th International Conference on Control and Automation, Hungarian Acad Sci, Budapest, Hungary, 993-996.
14. Sharifalundian, E.: Compression of vital bio-signals of patient by enhanced set partitioning in hierarchical trees algorithm, World Congress on Medical Physics and Medical Bioengineering, (2009).
15. Perner P. (2008). Prototype-based classification. Applied Intelligence 28(3): 238-246.
16. Perner P. (2006). Case-base maintenance by conceptual clustering of graphs. Engineering Applications of Artificial Intelligence, Vol. 19, No. 4, pp. 381-295.
17. Ciufudean C, Filote C. (2005). Performance evaluation of distributed systems, international conference on control and automation. ICCA 2005, June 26-29, Budapest, Hungary, ISBN0-7803-9138-1, IEEE Catalogue Number: 05EX1076C, pp.21-25.
18. Leão CP, Soares FO, Machado JM, Seabra E, Rodrigues H. (2011). Design and development of an industrial network laboratory. International Journal of Emerging Technologies in Learning, Vol. 6, Special Issue 2, pp. 21-26.
19. Barros C, Leão CP, Soares F, Minas G, Machado J. (2013). RePhyS: A multidisciplinary experience in remote physiological systems laboratory. International Journal of Online Engineering, Vol. 9, Special Issue 2, pp. 21-24.
20. Kunz G, Machado J, Perondi E. (2015). Using timed automata for modelling, simulating and verifying networked systems controller's specifications. Neural Computing and Applications, pp. 1-11. Article in Press.

Business Survey OMR Sheet Scanning and Reading Using Webcam as OMR Reader

Shashank Shekhar[1], Kushal Lokhande[1], Sushanta Mishra[1],
Sam Pritchett[1], Parag Chitalia[1], Akash[2]

[1]VMware Inc.
{sshekhar, klokhande, skumarmishra, spritchett, pchitalia}@vmware.com
[2]cmaster.akash@gmail.com

Abstract. Optical Mark Reader (OMR), also called "mark sensing", is a method of scanning technology in which data is input into a computer system via marks made in predefined positions on a form. Therefore, OMR is best for handling discrete data. A considerable number of VMware partners and customers record their business surveys on sheets similar to OMR sheets (especially during hands on lab sessions and conferences) and the digitization of these survey OMR sheets takes a lot of manual effort or require costly OMR Scanners for reading them. Majority of partners and customers who are in B2B domain upload these OMR sheets on their FTP sites for future usage and reference. This paper is an outcome of trying to tackle the problem of digitization of these OMR sheets. It came out to be a challenging task as this technique was required to be precise and efficient enough to replace the existing costly but highly accurate OMR Scanners. This technique has been designed and implemented with the help of open source computer vision library, viz, OpenCV and image processing methods.

Keywords: OMR, Mark sensing, OpenCV, image processing.

1. Introduction

A considerable number of VMware partners and customers record their business surveys on sheets similar to OMR sheets and the digitization of these survey OMR sheets takes a lot of manual effort or require costly OMR Scanners for reading them. Our technique of OMR scanning focuses on completely removing the currently present OMR Scanners in an efficient manner by making use of a video camera as an input device for scanning OMR sheet for digitization and then evaluating the OMR.

An OMR sheet is placed in front of the video camera and a program is made to run through Open source computer vision libraries implemented on C programming platform. The camera acts as an input device that captures the image of an OMR sheet placed in front of it and then that image is processed by the system through a program that follows a series of steps specified in the paper which produces the output telling which mark is filled-in at what position and thus we find the answers as per our requirement. This technique mainly focuses on reducing the cost of ownership of costly OMR Scanners.

We provide functions necessary to run the webcam and use a captured image after saving it and loading it for processing using open source computer vision library. This paper has four sections, the first one being Introduction. Introduction is followed by the second section, which gives an overview of Open Source Computer Vision (OpenCV) Library. The third section provides an explanation of the image thresholding. The fourth section is about image gridding and image division. The fifth part is about output generation using the processed images. The last section, which is section six, concludes the work, discussing the future scope.

2. Solution Approach

2.1. OpenCV Library

OpenCV (Open Source Computer Vision Library) is a library of programming functions mainly aimed at real time computer vision, developed by Intel [1]. OpenCV is written in C++ and its primary interface is in C++. OpenCV runs on Windows, Android, Linux and OS X. The user can get official releases from Source Forge. OpenCV runs real time and the libraries present in OpenCV were linked to the Borland Dev Cpp. There are predefined functions present in OpenCV [2] which helped us in running the webcam and saving the images taken as input by the webcam at real time. We have made use of OpenCV to create software that takes image of an OMR sheet and our technique makes it accept only the region containing the questions for processing. The technique has made use of image thresholding which has been discussed in the next section.

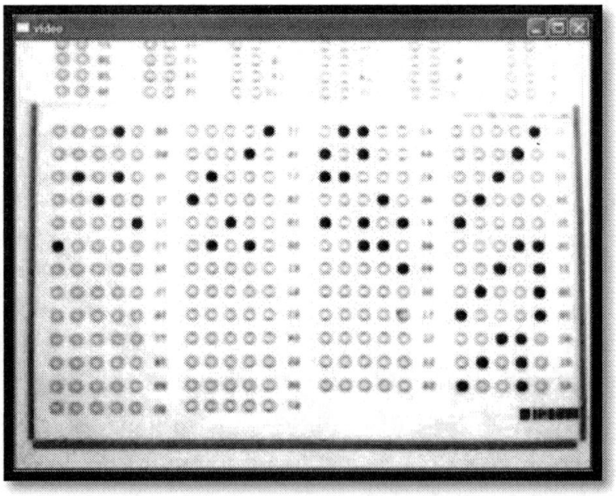

Fig. 2.1. OMR Sheet reading through webcam using OpenCV.

The OMR Sheet is saved using OpenCV library functions after the library is linked to Borland Dev Cpp compiler. This technique is developed on Windows OS.

2.2. Image Thresholding

Thresholding is the simplest form of image segmentation and is used to create binary images. The images are in black and white. The black part in a thresholded image is undetected part and the white part is detected part. The pink part is made by us and it has been made outside the print of OMR sheet without tampering with the print. Figure (**2.1**) is thresholded and processed using OpenCV to form the white part or detected part in the binary thresholded image. The detected part is then used for further application. Figure (**2.2**) shows a binary thresholded image of the detected part which we have made use of in our software.

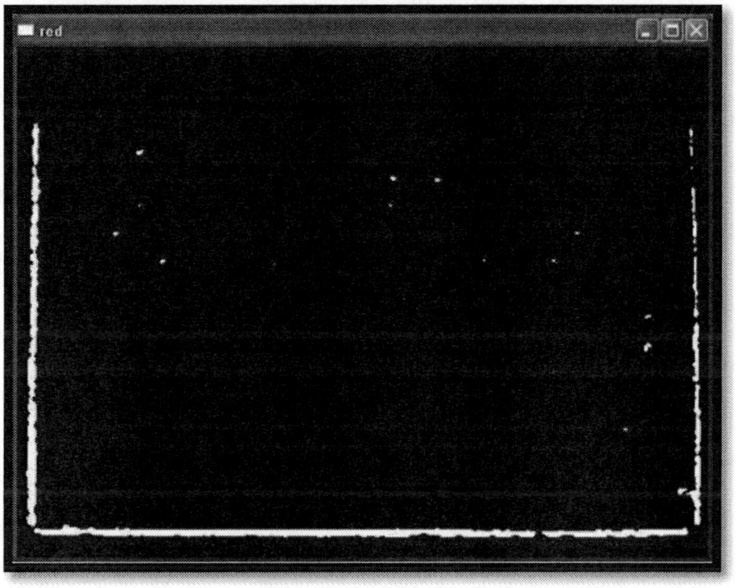

Fig. 2.2. Thresholded image showing white detected part for processing.

Next, we find a bounding rectangle for the thresholded part in the image [3]. That bounding rectangle contained all the black marks, which are answers to the questions. These marks are read by detection of black color and the same color is thresholded. We can find the location of the mark as per our requirement after the marks are detected. Figure **2.3** shows white colored detected marks and the rest image is undetected.

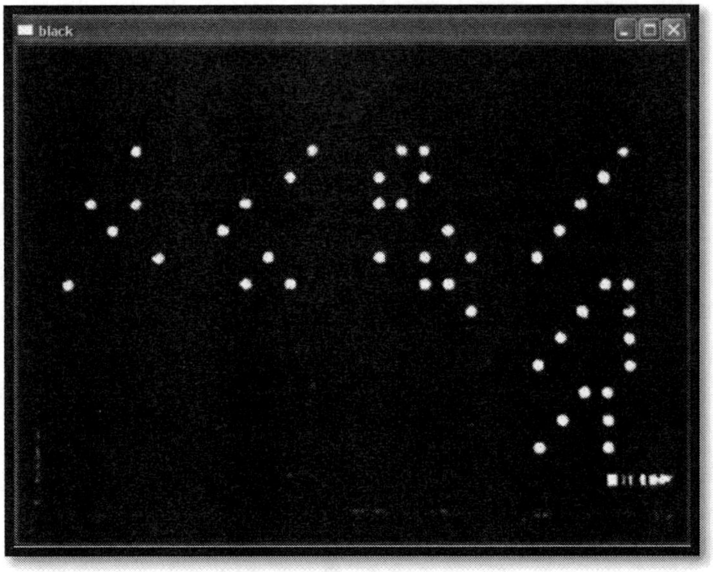

Fig. 2.3. Thresholded image showing white detected marks for processing.

Thresholding is used to differentiate between the desired and undesired portions in an image. In the later section, image gridding and division, explains how each mark is treated as a separate image and processed. Thresholding tells us that the white region is detected successfully and can now be processed as required by the programmer. We can calculate area, orientation, height, length, breadth and many other characteristics of thresholded part in the image. The thresholding is vision based, we can use real time input of multiple OMR Sheets and Optical Marks related to the image.

2.3. Image Gridding and Division

Image gridding involves making a grid as desired on the image. The grid will be drawn over the image in such a manner that each square or rectangle in the grid contains an optical mark or the black dot. After this, we differentiate between each black dot according the rectangle in which it is contained. The images are divided into two-dimensional matrix using array containing the row and column number of the rectangle telling whether it is filled by a black dot [4].

Firstly, we draw a grid over the image, which is shown, on the screen in a separate window allotted to the image. A separate image is used to draw only the grid, which is overlapped over the OMR Sheet. The grid is made to work even if the sheet is not at a fixed distance from the webcam. According the ratio of the distance, the grid adjusts itself. Figure **2.4** shows grid drawn using yellow color over the input image. Each rectangle in the grid contains a black dot.

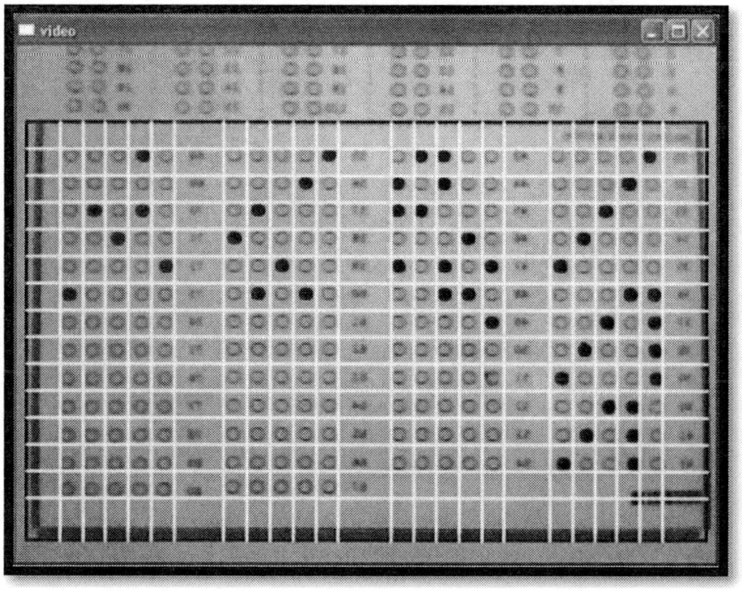

Fig. 2.4. Custom made image-containing grid overlapped over OMR Sheet.

As you can see that, each rectangle in the image contains a black mark or a black dot. With the help of the grid, we can differentiate between all the rectangles in the image. Now the grid is used to break the problem into four major rectangles, which can be processed separately. The first two columns contain thirteen questions and the last two contain twelve questions, which makes it difficult to find a single large rectangle and then process it. Four different rectangles each for separate set of questions make it easier to keep track of the small rectangles containing the black marks and thus it can be easily solved.

Figure (**2.5**) shows the four separate rectangles provided to each set of questions according the column in which they are placed. We can also see that the columns are not equally spaced which is another reason why we need to separate the set of questions. In the following figure, we notice that all the questions are not covered because we have emphasized on solving all the possibilities of answering the OMR Sheet by a candidate.

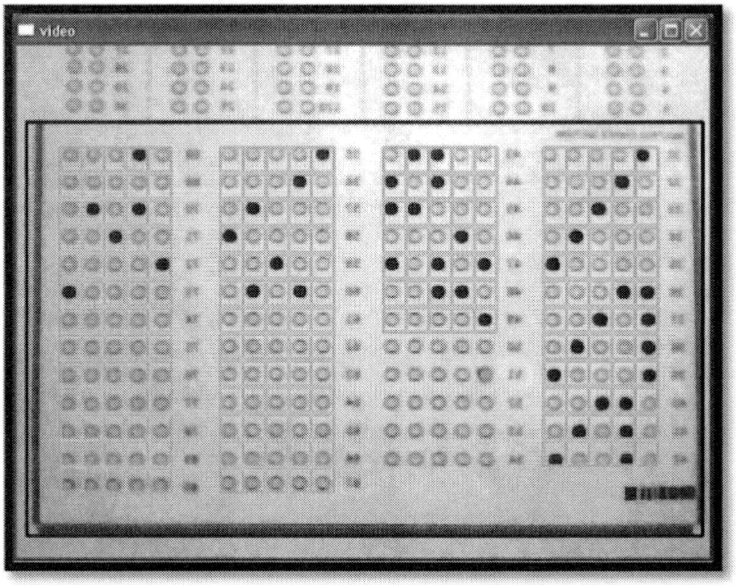

Fig. 2.5. Division of questions according the rectangle in which they are contained.

Now that we have the containing rectangle of each of the black mark we will use IMAGE DIVISION to divide the image into separate rectangles for each mark. If we treat the first rectangle and traverse it row wise we will first divide for all columns of first row, then second, then third and so on till the whole rectangle is divided. Figure (**2.6**) shows the division of each black mark to a separate image shown in separate window allocated to the image.

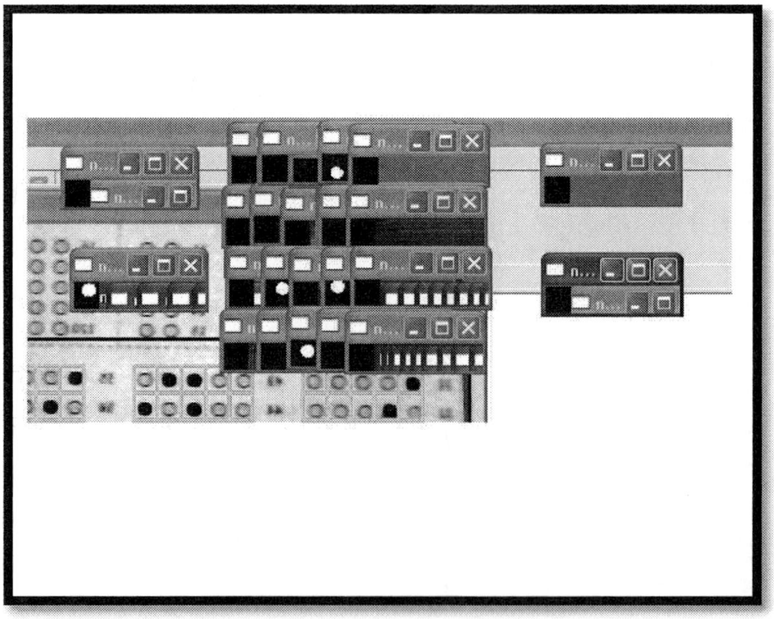

Fig. 2.6. Division of black marks to separate images shown in separate window.

The group of images are corresponding to first few questions of the first rectangle. The images are named as n[row][column]. The first row is zero row and columns are from zero to four. Accordingly, all the rectangles are divided. Working on so many images is supposed to make the software heavy and difficult to manage but OpenCV provides with functions to decide whether to show, hide, create and release these images after they serve their purpose.

The images created in array are fed with complete black image and the small rectangles Figure (**2.6**) are used to set only that portion of the thresholded image recognizing the black marks as the region of interest using function provided by OpenCV library. Then that region of interest is set as an image and shown in the separate window named according the order in which the rectangle is present to avoid confusion.

3. Utilizing Output

The output generated tells if the question is filled in a valid manner or not. If invalid, further processing of that question does not take place. If a question is validly filled then we check which part of the image has a black mark, as only one image belonging to a row will contain a mark. We calculate area of the mark and have confidence level of point nine, which will provide accurate judgement. If we come across a value, that is less than that of the estimated, the mark is considered to be partially filled and in turn, this will be considered as a wrong answer. In the above manner, we find all the answers and solve the OMR Sheet digitization problem by making use of just a common webcam and OpenCV library. The values change according ratio of the distance of the OMR Sheet to the webcam and the area calculated of the mark.

The output is stored in an array. Figure **3** shows the output telling which mark is filled where. The programmer can use any form of technique to provide answer as required by her.

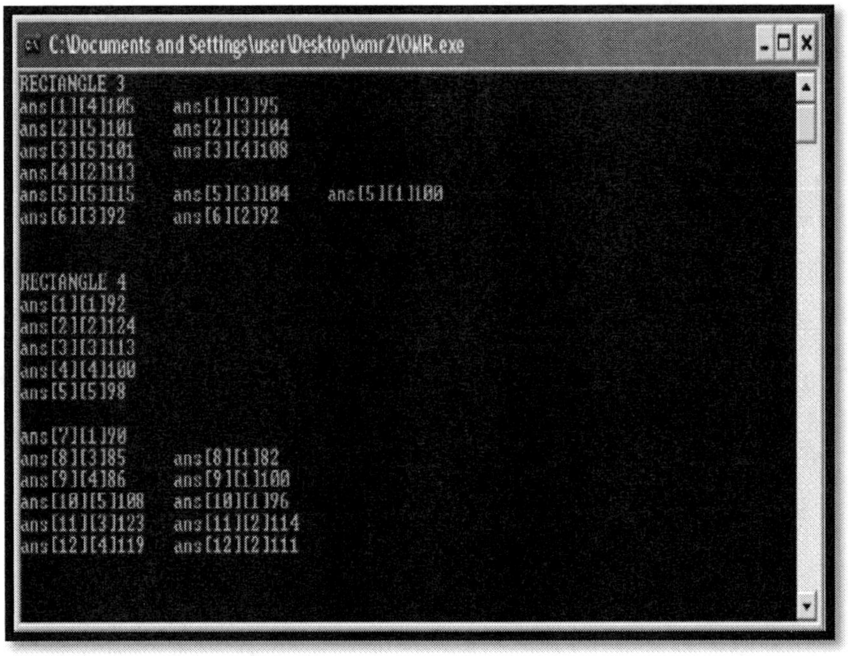

Fig. 3. Output window showing the answers in the form of array in which they are stored.

4. Conclusion

This solution helped us read survey responses listed on partners' websites accurately and enabled us to analyze the responses. Using our technique, we have been able to bring in efficiency to a critical business process. We see that by installing Open Source Computer Vision library and a webcam we can replace the OMR Scanners and reduce the cost of ownership to near nil. It also helps in reducing cost of training personnel and maintenance of machines as our technique is very accurate and fast. We have started working on taking multiple images at a time, which will help in even faster processing and save time. Our technique can be implemented in solving similar problems where OMR Sheet reading and understanding is required. The code for the application can be transferred to a Raspberry PI or similar device, having a camera attached to it to make an independent device out of the concept we have developed.

References

1. OpenCV Library: http://www.sourceforge.net/projects/opencvlibrary
2. Description of OpenCV functions and usage: http://docs.opencv.org/3.0-beta/doc/tutorials/tutorials.html
3. Thresholding function and usage: http://opencv-python-tutroals.readthedocs.io/en/latest/py_tutorials/py_imgproc/py_thresholding/py_thresholding.html

4. García GB, Suarez OD, Aranda JLE, Tercero JS, Gracia IS, Enano NV. (March 2015). Learning Image Processing with OpenCV.
5. Broeke J, Pérez JMM, Pascau J. (November 2015). Image Processing with ImageJ - Second Edition.
6. Kaehler A, Bradski G. (December 2016). Learning OpenCV 3, Computer Vision in C++ with the OpenCV Library.

Modelling Resistive Random Access Memories by Means of Functional Principal Component Analysis

A. M. Aguilera[1], M.C. Aguilera-Morillo[2], F. Jiménez-Molinos[3], and J.B. Roldán[3]

[1]FDA of RRAMs Aguilera et al. Departamento de Estadística. e I.O. Universidad de Granada, Spain
aaguiler@ugr.es WWW home page: https://sites.google.com/site/amiaguilerapino/
[2]Departamento de Estadística. Universidad Carlos III of Madrid, Spain
[3]Departamento de Electrónica y Tecnología de los Computadores. Universidad de Granada, Spain

Abstract. Resistive Random Access Memories (RRAMs), based on the resistive switching of transition metal oxide films (TMOs), is one of the strong candidates for future nonvolatile applications due to their good scalability, long endurance, fast switching speed, and ease of integration in the back end of the line of CMOS processing. The switching mechanism consists of an initial high resistance state (HRS) of the device that converts into a low resistance state (LRS) by applying voltage in the set process. Succeeding application of voltage returns the HRS in a reset process. In practice, this leads to development and breakdown of a conductive filament (CF) and provides thousands of current-voltage curves observed at a big and dense set of voltages. Functional data analysis (FDA) methodologies are applied in this paper for modelling and explaining the variability associated with the stochastic process generating this big data set. The reduction dimension technique Functional Principal Component Analysis (FPCA) provides a simple model that explains the majority variability in terms of only one scalar variable highly correlated with the voltage to reset.

Keywords: Functional data analysis, penalized splines, functional principal components, resistive memories, switching process.

1. Introduction

In the current electronics industry landscape scaling issues constitute the major driving force for advance. In the last decades huge efforts have been made to go beyond the hurdles found in the scaling process. In general, scaling allows integrating more electronics devices in the same silicon chip and this means a cost reduction as well as an enhancement of the operation speed and storage capacitance of the circuits integrated in a chip. In this context, new devices come into play that allow the increase of integration. In particular, in the field of non-volatile memories, a lot of money is being invested to keep up the pace of other electronics circuits such as microprocessors and volatile memories. The current technological momentum of non-volatile memory chips is dominated by floating-gate transistors [1-3] although a promising group of new devices is being deeply analyzed. Among these devices, the most interesting are the following: resistive switching memories (RRAMs) [1-5], phase change memories (PCM) [6] and spin-transfer torque RAMs (STT-RAM) [7].

RRAMs are devices with a very simple physical structure. Two metal plates acting as electrodes with a dielectric in between. Therefore, their scaling possibilities are amazing taking into consideration fabrication techniques such as atomic layer deposition (ALD). In addition, the resistive state can be easily switched in a non-volatile and swift manner; moreover, other features such as endurance, low power operation and compatibility with current CMOS technology [4, 8], make them serious candidates to substitute the current floating gate transistor technology in future NVM chips [1-3].

Apart from experimental research, a step forward in the field of compact modeling is needed to lead RRAM technology to maturity at the industry standards. Compact models describe by means of analytical expressions the electric currents, capacitances and other measurable magnitudes by employing the voltages applied at their terminals as independent variables. These compact models are introduced in circuit simulators, using Kirchhoff's laws, to design electronic circuits. Therefore, a set of analytical equations, usually explicit, and the corresponding fitting parameters characterize a determined RRAM technology. Although different compact models for RRAMs can be found in the literature [9-14], there are many open issues. In this respect, the fact that the physical mechanisms behind resistive switching, the core of RRAM operation, are stochastic [3, 5] present important numerical problems for the correct modeling. The conduction of RRAM is filamentary in many cases; this implies that conductive filaments are formed and destroyed and because of this, the device resistance switches from a High Resistance State (HRS) to a Low Resistance State (LRS). A set process lets the device switch from HRS to LRS, and inversely the reset process change the device resistive states the other way around.

The filaments are made of random clustering of metallic ions or oxygen vacancies [2, 3, 5, 15-17]. The stochastic nature of the device operation makes essential the use of advanced mathematical techniques to deepen in the RRAM compact modeling territory; that is where Functional Data Analysis (FDA) comes into play. The name FDA agglutinates a set of statistical methods for the analysis of samples of curves or more general functions [23, 25, 34]. The powerful tools that can be developed based on FDA procedures can be used to attack many real world sophisticated problems [24], among them the coherent modeling of the physics behind RRAM technology. The basis tool in FDA is the reduction dimension technique Functional Principal Component Analysis [30 31, 35-37]. In this contribution a new RRAM modeling method is introduced based on FPCA and applied to experimental current curves measured in state-of-the-art devices as well as to simulated current curves. An alternative method for analyzing the main features of the sample curves associated with RRAMs could be using Delta-Modulation and Similarity as shown in [41] with spectrometric curves.

Another approach employed in the electronics industry to help in the technology development process is linked to use of device simulators. There are huge commercial simulation suites that can be of great help in all the development steps; nevertheless, for new devices the first simulation tools are generated at research level, that is the case of RRAMs [3, 5, 15, 17-21]. Simulators solve the differential equations that describe the physics in detailed simulation domains and allow the knowledge of internal variables such as temperature, electrical conductivity, *etc*., that can be very useful. We have employed simulation results to improve the accuracy of our compact model.

The contents of this contribution are sorted out in several sections. In Section 2, the technological details of the RRAMs and the simulator that was previously developed are given. In Section 3 the main features of the modeling procedure based on FPCA are explained. Section 4 is devoted to discussion of the main results and future research.

2. Device Fabrication and Simulator Description

The devices studied here were designed and fabricated at the Institut of Microelectronics of Barcelona IMB-CNM (CSIC), they are based on a Ni/$\mathrm{Hf}O_2/Si-n^+$ stack. The ALD fabricated dielectric layer was 20nm thick; further details of the fabrication process and measurement setup can be found in [22]. The conduction is filamentary in these devices, *i.e.*, it takes place trough conductive filaments (CFs) that are formed and destroyed within the RS device operation [15, 22].

A few experimental current-voltage curves corresponding to several RS (set-reset) cycles are shown in Figure **1**. Reset curves are shown in blue while set curves are plotted in green. The curves correspond to different cycles from a set-reset series of almost three thousands cycles. The curves are different in all the cases due to the stochastic nature of the processes behind the conductive filament formation that fix the device resistance, *i.e.*, the ratio between the device voltage and the corresponding current [1, 3, 5, 22].

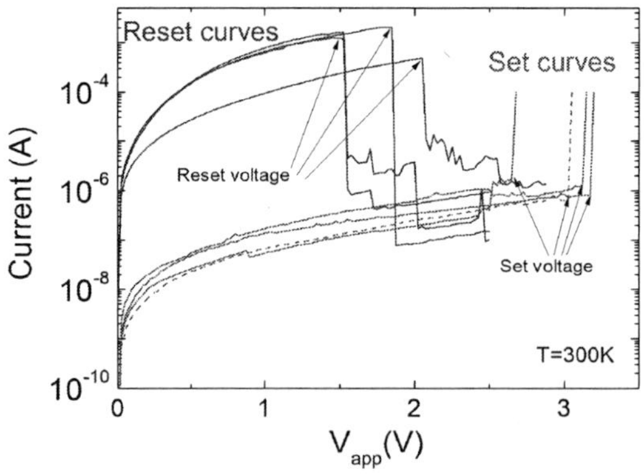

Fig. 1. Current versus applied voltage for the fabricated devices reported above. Several set/reset cycles in a long RS series for devices based on a Ni/$\mathrm{Hf}O_2/Si-n^+$ stack are included. Although the Ni electrode had a negative voltage applied while the substrate was grounded [22], we have considered absolute values for the applied voltage for the sake of simplicity.The V_{Reset} and V_{Set} points are highlighted.

It can be observed in Figure **1** that the reset point is determined by the sudden drop of the current; the rupture of the conductive filament [17, 22] lies behind this current drop. At this point the device reaches the HRS. The V_{Set} points are also marked, these points are characterized by the creation of the conductive filament and the LRS is established after the set process. The filament formation makes the RRAM resistance greatly diminish.

For the simulator employed here to describe the RRAMs reported above, an arbitrary number of conductive filaments with truncated-cone shapes and electrically coupled were employed. We changed the filaments initial radius, including the series and Maxwell resistances for each filament [15, 20, 21]. Quantum effects were included by the inclusion of the Quantum Point Contact model in the current calculation. The conductive filament dynamics was considered solving the chemical redox equations controlling the CF radius evolution along with the diffusion of metallic species; the temperature distribution was obtained by solving the heat equation with the current and redox equations in a self-consistent numerical scheme [15, 20, 21]. The conductive filament shape is shown in Figure **2**.

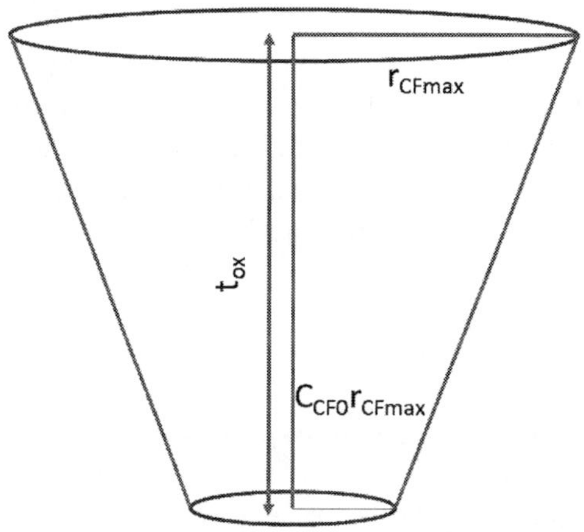

Fig. 2. RRAM scheme employed in the simulations. The conductive filament formed in the dielectric is represented with a truncated-cone shape. The maximum radius is r_{CFmax}, the lower radius is represented as a fraction of the maximum one $C_{CF0}r_{CFmax}$, where C_{CF0} value is between 0 and 1. The dielectric thickness is assumed to be t_{ox}.

We have performed several simulations for RRAM with a single conductive filament of different sizes (see Figure **3**). The shape of the filament was a truncated-cone with the maximum radius indicated in the figure and a minimum radius corresponding to a third of the maximum radius in all the cases. The simulated structure was the one reported above for the fabricated devices including a dielectric thickness of 20nm. See the influence of the CF size in the final device current.

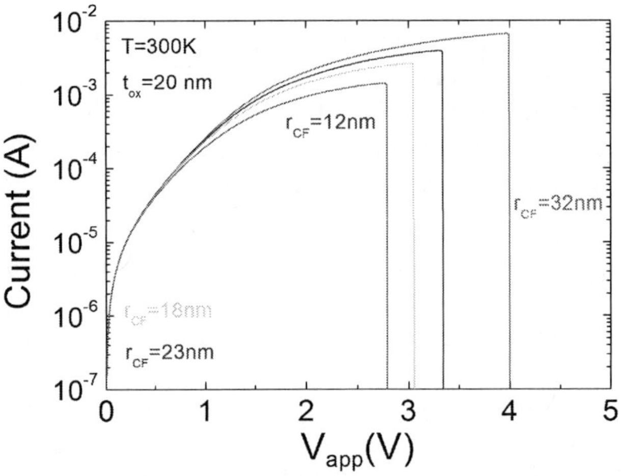

Fig. 3. Current versus applied voltage for the simulated devices reported above. Several reset cycles for devices based on a Ni/Hf$O_2/Si - n^+$ stack with conductive filaments of different sizes are considered.

As it will be shown in the next section, the reset voltage (V_{reset}) is a key parameter for our modeling procedure, since all the reset curves studied are defined between zero and the corresponding V_{reset}. Once all the curves in the simulated or measured series are normalized in a common interval, other FDA steps as basis representation and FPCA are performed to complete the model.

3. Functional Principal Component Analysis

As stated before, FPCA is applied for explaining the stochastic variability of the curves of evolution of electric current (I) in terms of voltage (V) from a sample of reset cycles in Resistive Random Access Memories (RRAMs).

Let us denote by $\{I(v): v \in T\}$ the stochastic process that represents current in terms of voltage in a real interval T. And let us suppose that $\{I(v)\}$ is a second order and continuous in quadratic mean stochastic process whose sample functions belong to the Hilbert space $L^2[T]$ of square integrable functions on T.

Functional FPCA is a reduction dimension technique that was introduced by [30] as a generalization of multivariate principal component analysis to the case of a continuous-time stochastic processes. The *j-th* functional principal component score is given by

$$\xi_j = \int_T (I(v) - \mu(v)) f_j(v) dv,$$

where the weight function or loading f_j is obtained by solving

$$\begin{cases} Max_f \quad Var\left[\int_T (I(v) - \mu(v)) f(v) dv\right] \\ r.t. \quad \| f \|^2 = 1 \ and \ \int f_\ell(v) f(v) dv = 0, \quad \ell = 1, \ldots, j-1. \end{cases}$$

It can be shown that the weight functions are the eigenfunctions of the covariance opcrator of I. That is, the solutions to the eigenequation

$$\int_T C(v,u)f_j(u)du = \lambda_j f_j(v), \quad v \in T,$$

so that the principal components $\{\xi_j\}$ are uncorrelated zero-mean random variables with maximum variance given by $Var(\xi_j) = \lambda_j$.

Then, the following orthogonal decomposition of the process in terms of uncorrelated random variables and deterministic functions is obtained:

$$I(v) = \mu(v) + \sum_{j=1}^{\infty} f_j(v)\xi_j. \tag{1}$$

This expression is a probabilistic result known as Karhunen-Loève orthogonal expansion. In practice, the series (1) is truncated in term of the first q principal components so that the variance explained by them ($\sum_{j=1}^{q} \lambda_j / \sum_{j=1}^{\infty} \lambda_j$) is as close as possible to one.

3.1. Estimation Algorithm

In practice, to estimate functional principal component analysis associated to RRAMs we have a random sample of n reset curves denoted by

$$\{I_i(v): i = 1, \ldots, n; v \in [0, V_{i-reset}]\},$$

where $V_{i-reset}$ is the voltage to reset. The estimation of FPCA presents two important problems. On the one hand, the reset curves are not defined on the same domain because the voltage to reset is different among cycles. In addition, the data consist of discrete observations of each reset curve $I_i(v)$ at a finite set of current values until the $V_{i-reset}$ point.

These problems are solved by the following three step estimation procedure:

1. Registration of the reset curves in the interval [0,1].

It will be do in the simplest way that consists of transforming the domain $[0, V_{i-reset}]$ of each reset curve in the interval [0,1] by the function $v/V_{i-reset}$. This way, the sampling points where the registered curves are observed are different in number and also in position.

2. Reconstruction of the reset curves by P-spline smoothing in terms of B-spline basis expansions.

In order to reconstruct the true form of reset curves the following basis representation is considered for the registered curves:

$$I_i^*(u) = \sum_{j=1}^{p} a_{ij}\phi_j(u), \; i = 1, \ldots, n, \tag{2}$$

in terms of a basis $\{\phi_1(t), \ldots, \phi_p(t)\}$.

Assuming that the reset curves are smooth and observed with error, penalized least squares approximation with B-splines basis is used to approximate the basis coefficients a_{ij}. The main advantages of using P-splines are that the choice and position of knots is not determinant so that it is sufficient to choose a relatively large number of equally spaced basis knots [38]. On the other hand, they have less numerical complexity and computational cost.

3. Functional PCA of the P-spline representation of reset curves.

Once the basis coefficients of the registered curves are estimated, FPCA of basis representation given by (2) is equivalent to multivariate PCA of matrix $A\Psi^{\frac{1}{2}}$, with $A = (a_{ij})_{n\times p}$ being the matrix of basis coefficients and $\Psi^{\frac{1}{2}}$ being the squared root of the matrix of inner products between basis functions functions [37].

4. First Results and Open Lines of Research

The ideas presented in this work are part of a new research line whose main objective is connected with the application of functional data analysis methodologies to model stochastic characteristics in the switching processes associated with the operation of RRAMs. Actually, the FPCA approach described before has been performed on a sample of reset cycles simulated with different truncated-cone shaped filament dimensions, the structures that produce resistive switching in RRAMs.

The obtained results provide that only one principal component explains more that a 95% of the total variability of the registered reset process. This fact means that a simple model, with only one scalar variable, can be used to simulate these processes as $I^*(v) = \mu(v) + \xi_1 f_1(v)$ in the interval [0,1]. The results show promising developments both from the characterization and from the modeling viewpoints. In fact, the scalar variable extracted with the new numerical algorithm is highly correlated with the threshold voltage, V_{reset}, so that the corresponding probability distribution could provide a complete modeling description of the reset process.

The next step in this study is linked to the estimation of a model making use of a big sample of real curves. This would be achieved by employing functional regression models [33, 39, 40] to explain the main reset process characteristics in terms of different electrode metals and conductive filament geometries.

Acknowledgements

We thank the Spanish Ministry of Economy and Competitiveness for Projects MTM2013-47929-P and TEC2014-52152-C3-2-R (also supported by the FEDER program), and Consejería de Innovación, Ciencia y Empresa from the Junta de Andalucía, Spain, for project P11-FQM-8068. We would like also to thank F. Campabadal and M. B. González from the IMB-CNM (CSIC) in Barcelona for fabricating and providing the experimental measurements of the devices employed here.

References

1. XieY. (2014). Emerging Memory Technologies. Springer.
2. Lanza M. (2014). A review on resistive switching in high-k dielectrics: A nanoscale point of view using conductive atomic force microscope. Materials 7, 2155.
3. Pan F, Gao S, Chen C, Song C, Zeng F. (2014). Recent progress in resistive random access memories: materials, switching mechanisms and performance. Materials Science and Engineering, 24, 421.
4. Waser R, Aono M. (2007). Nanoionics-based resistive switching memories. Nature Materials, 6, 833-840.
5. Lee J, Lee S, Noh T. (2015)Resistive switching phenomena: A review of statistical physics approaches. Applied Physics Reviews, 2, 031303.
6. Wong HSP, Raoux S, Kim S, Liang J, Reifenberg J, Rajendran B, Asheghi M, Goodson K. (2010). Phase change memory. Proceedings of The IEEE – PIEEE, 98, 2201-2227.
7. Zhao W, Chaudhuri S, Accoto C, Klein JO, Ravelosona D, Chappert C, Mazoyer P. (2012). High density spin-transfer torque (SST)-MRAM based on cross-point architecture. 4th IEEE International Memory Workshop (IMW), 1-4.
8. Strukov DB, Snider GS, Stewart DR, Williams RS. (2008). The missing memristor found. Nature 453, 80-83.
9. Picos R, Roldan JB, Al Chawa MN, Jimenez-Molinos F, Villena MA, Garcia-Moreno E. (2009). Exploring ReRAM-based memristors in the charge-flux domain, a modeling approach. Proceedings of International Conference on Memristive Systems, MEMRISYS2015.
10. Biolek Z, Biolek D, Biolkova V. (2009). Spice model of memristor with nonlinear dopant drift. Radioengineering, 18, 210-214.
11. Jimenez-Molinos F, Villena MA, Roldan JB, Roldan AM. (2015). A SPICE compact model for unipolar RRAM reset process analysis. IEEE Transactions on Electron Devices, 62, 955-962.
12. Shin S, Kim K, Kang SM. (2010). Compact models for memristors based on charge-flux constitutive relationships. Computer-Aided Design of Integrated Circuits and Systems, 29, 590-598.
13. Gonzalez-Cordero G, Jimenez-Molinos F, Roldan JB, Gonzalez M, Campabadal F. (2017). An in-depth study of the physics behind resistive switching in TiN/Ti/HfO2/W structures. Journal of Vacuum Science and Technology, 35, 01A110.
14. Gonzalez-Cordero G, Roldan JB, Jimenez-Molinos F, Sune J, Long S, Liu M. (2016). A new model for bipolar rrams based on truncated cone conductive filaments, a Verilog-A approach. Semiconductor Science and Technology, 31, 115013.
15. Villena MA, Gonzalez MB, Jimenez-Molinos F, Campabadal F, Roldan JB, Sune J, Romera E, Miranda E. (2014). Simulation of thermal reset transitions in resistive switching memories including quantum effects. Journal of Applied Physics 115, 214504.
16. Larentis S, Nardi F, Balatti S, Gilmer DC, Ielmini D. (2012). Resistive switching by voltage-driven ion migration in bipolar RRAM. part II:Modeling. IEEE

Transactions on Electron Devices, 59, 2468-2475.
17. Villena MA, Jimenez-Molinos F, Roldan JB, Sune J, Long S, Lian X, Gamiz F, Liu M. (2013). An in-depth simulation study of thermal reset transitions in resistive switching memories. Journal of Applied Physics, 114, 144505.
18. Degraeve R, Fantini A, Raghavan N, Goux L, Clima S, Chen YY, Belmonte A, Cosemans S, Govoreanu A, Wouters DJ, Roussel Ph, Kar, GS, Groeseneken G, Jurczak M. (2014). Hourglass concept for RRAM: A dynamic and statistical device model. in Proceedings of the 21th International Symposium on the Physical and Failure Analysis of Integrated Circuits (IPFA).
19. Villena MA, Roldan JB, Jimenez-Molinos F, Sune J, Long S, Miranda E, Liu M. (2014). A comprehensive analysis on progressive reset transitions in RRAMs. Journal of Physics D: Applied Physics, 47, 205102.
20. Villena MA, Gonzalez MB, Roldan JB, Campabadal F, Jimenez-Molinos F, Gomez-Campos F, Sune J. (2015). An in-depth study of thermal effects in reset transitions in HfO2 based RRAMs. Solid-State Electronics, 111, 47-51.
21. Villena MA, Roldan JB, Gonzalez MB, Gonzalez-Rodelas P, Jimenez-Molinos F, Campabadal F, Barrera D. (2016). A new parameter to characterize the charge transport regime in Ni/HfO2/Si-n+-based RRAMs. Solid-State Electronics, 118, 56-60.
22. Gonzalez MB, Rafi J, Beldarrain O, Zabala M, Campabadal F. (2014). Analysis of the switching variability in Ni/HfO2-based RRAM devices. IEEE Transactions on Device and Materials Reliability, 14, 769-771.
23. Ramsay JO, Silverman BW. (2005). Functional data analysis (Second Edition). Springer-Verlag.
24. Ramsay JO, Silverman BW. (2002). Applied functional data analysis: Methods and case studies. Springer-Verlag
25. Horvath L, Kokoszka P. (2012). Inference for functional data with applications. Springer-Verlag.
26. Ramsay JO, Hooker G, Graves S. (2009). Functional Data Analysis with R and MATLAB. Springer-Verlag.
27. Aguilera AM, Aguilera-Morillo MC. (2013). Comparative study of different B-spline approaches for functional data. Mathematical and Computer Modelling, 58(7-8), 1568-1579.
28. Aguilera AM, Aguilera-Morillo MC. (2013). Penalized PCA approaches for B-spline expansions of smooth functional data. Applied Mathematics and Computation, 219(14), 7805-7819.
29. Aguilera-Morillo MC, Aguilera AM, Escabias M, Valderrama MJ. (2013). Penalized spline approaches for functional logit regression. Test, 22(2), 251-277.
30. Deville JC. (1974). Méthodes statistiques et numÂ´ériques de l'analyse harmonique. Annales *de l'INSEE,*15, 3-101.
31. Aguilera AM, Gutiérrez R, Valderrama MJ. (1996). Approximation of estimators in the PCA of a stochastic proces using B-splines. Communications in Statistics. Simulation and Computation, 25(3), 671-690
32. Eilers PHC, Marx BD. (1996). Flexible smoothing with B-splines and penalties. Statistical Science, 11(2), 89-121.
33. Aguilera AM, Aguilera-Morillo MC, Preda C. (2016)Penalized versions of functional pls regression. Chemometrics and Intelligent Laboratory Systems, 154,

80-92.
34. Zhang, J-T. (2014). Analysis of Variance for functional data. CRC Press.
35. Dauxois J, Pousse A, Romain Y. (1982). Asymptotic theory for the principal component analysis of a vector random function: some applications to statistical inference. Journal of Multivariate Analysis, 12(1), 136-156.
36. OcaËœña FA, Aguilera AM, Valderrama MJ. (1999). Functional principal components analysis by choice of norm. Journal of Multivariate Analysis, 71, 262-276.
37. OcaËœña FA, Aguilera AM, Escabias M. (2007). Computational considerations in functional principal component analysis. Computational Statistics, 22(3), 449-465.
38. Ruppert D. (2002). Selecting the number of knots for penalized splines. Journal of Computational and Graphical Statistics 11, 735-757.
39. Escabias M, Aguilera AM, Valderrama MJ. (2004). Principal component estimation of functional logistic regression: discussion of two different approaches. Journal of Nonparametric Statistics, 16, 365-384.
40. Aguilera AM, Escabias M, OcaËœña FA, Valderrama MJ. (2015). Functional wavelet-based modelling of dependence detween dupus and dtress. Methodology and Computing in Applied Probability, 17(4), 1015-1028.
41. Perner P, Attig A, Machno O. (2011). A novel method for the interpretation of spectrometer signals based on delta-modulation and similarity. Transactions on Mass-Data Analysis of Images and Signals, 3(1), 3-14. Determination.

A Literature Survey on Handwriting Recognition: Trends and Methodologies Handwriting Recognition

David Hoffman[1] and Nasseh Tabrizi[2]

[1]East Carolina University
Department of Computer Science
Science and Technology Building
Greenville, NC 27858-4353 USA
Phone: +1-919-675-2393 cell
hoffmand12@students.ecu.edu
[2]East Carolina University
Department of Computer Science
Science and Technology Building
Greenville, NC 27858-4353 USA
Phone: +1-252-328-9691 office
tabrizim@ecu.edu

Abstract. Handwriting recognition (HWR) involves computer transcription of handwritten documents. Researchers have been studying many techniques for HWR and have been more successful as computing power has increased and new algorithms have emerged. This paper presents the results of our survey of 141 articles from 38 journals conducted in this domain over the past ten years. We have tabulated articles by year of publication, journal, author, online vs. offline, script/language, dataset, methodology, and other special topics. Our research reveals that the Mixed National Institute of Standards and Technology Database (MNIST) is the most popular testing and training dataset for numerals and the Chinese Academy of Sciences' Institute of Automation (CASIA) is the next most popular dataset. Additionally, Arabic is the most popular language studied and *Pattern Recognition* is the most popular journal for HWR articles followed by *Pattern Recognition Letters*. Furthermore, there is a near even spread between online and offline techniques used in HWR with Hidden Markov Model being the most utilized methodology. The authors believe that the results reported in this paper will provide researchers and industry professionals with perception, future guidance, and recommendation on HWR research.

Keywords: Image processing, handwriting recognition, big data, survey.

1. Introduction

Handwriting recognition (HWR) is an area of computer image processing and pattern recognition in which handwritten texts are transcribed. Due to large volumes of handwritten documents, there is a need for machines to be able to recognize and transcribe historical documents [16, 17] bank checks [19], and postal addresses [41].

Handwriting recognition can be categorized into on-line and offline recognition. On-line recognition involves typically the use of a stylus input device in tandem with a tablet giving time and position data. In contrast, offline recognition involves analysis of documents.

In addition, many methods and algorithms have been applied to handwriting recognition. Common methodologies include:

1. sliding window approach coupled with Hidden Markov Models (HMM) [21, 37, 5]
2. holistic word analysis [29, 3]
3. character segmentation [28, 12]

The HMM method is currently very popular and has helped to produce high recognition rates. Holistic word analysis, in contrast to the sliding window, is less common but still prominent. Character segmentation of handwriting is slightly outdated; difficulties are caused by varying character widths, varying spacing, touching characters, *etc.* Character segmentation is, however, good for Optical Character Recognition (OCR) of typed documents.

Research in the field of HWR continues to progress, and recognition accuracies are improving. It is necessary for past articles to be reviewed and surveyed from time to time, paying particular attention to the feasibility of methods for the next generation of handwriting recognition.

The purpose of this study is to understand the trend of handwriting recognition research by examining literature, giving interested parties insight, and providing future direction on handwriting recognition. In this research survey, we have reviewed and classified articles on HWR that were published in journals between 2006 and 2015. Our research is organized as follows:

1. handwriting recognition background
2. related work
3. research methodology
4. classification process
5. results
6. conclusions, limitations, and implications

2. Handwriting Recognition Background

Handwriting recognition involves several steps in order to transcribe documents, as shown in Figure **1**. During pre-processing, the goal is to extract and normalize words such that they can be identified. Steps such as text line and word segmentation, followed by word normalization processes such as baseline estimation, slope/skew correction, slant correction, smoothing, and thinning are performed.

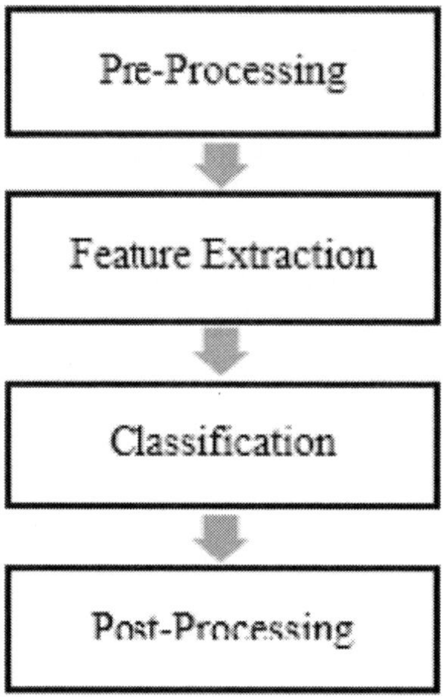

Fig. 1. Steps for handwriting recognition.

Once words have been segmented and normalized, features extraction techniques can be performed. During feature-extraction, a feature-vector is computed, giving shapes and features a numerical means of identification.

Classifiers, such as HMM [5, 21, 37] or Support Vector Machines (SVM) [4, 9, 25, 38, 39], are utilized next to process feature-vectors. Post-processing is completed to form and identify a word from a string of letters. Typically, the word is selected from a limited lexicon, an example being city names in the case of postal addresses. A smaller lexicon increases the possibility that the correct word is identified.

It is necessary to train and test these classification algorithms which is achieved by getting writing samples and manually associating the text transcription into the system or by using an existing dataset. There are several popular handwriting datasets against which authors have benchmarked their systems. The most popular dataset over the past 10 years is the Mixed National Institute of Standards and Technology Database of Handwritten Digits (MNIST) [26], which contains single numerical characters. Other common datasets, including those of words and sentences, are discussed in Section 5.6.

3. Related Work

Due to the vast amount of research and associated publications in this domain, several surveys on handwriting recognition have been published in journals and conference proceedings. HWR surveys include [1, 20, 23, 27, 31-33, 35, 36, 42-44].

This literature survey is different than other surveys aforementioned because we aim to report on the last ten years of publications. Rather than just comparing and contrasting various methods, our approach is to classify a selection of papers by year of publication, journal, author, online *vs.* offline, script/language, dataset, and methodology as in [34], a paper on the unrelated topic, recommender systems.

4. Research Methodology

A generic search for handwriting recognition on a library's academic article search tool yields 12,000+ results. Therefore we limited our search to specific databases, and executed the query: *title(handwriting recognition) OR title(cursive recognition).* As previously mentioned, we limited our results to journal articles within years 2006-2015. The 10-year period is considered to adequately cover current handwriting recognition research.

These electronic databases were searched with the query above to provide a broad bibliography of literature on handwriting recognition:

- IEEE Explore
- Proquest
- Science Direct

With regards to this survey, works not included are masters theses, doctoral dissertations, unpublished documents, non-English papers, news articles, and conference proceedings. Included are solely peer-reviewed journal papers. From the search results, we selected 141 papers providing a near-comprehensive review on research published in journal articles over the past 10 years.

5. Results

We classified articles by year of publication, journal, author, online *vs.* offline, script/language, dataset, methodology, and other special topics discussed in 5.8.

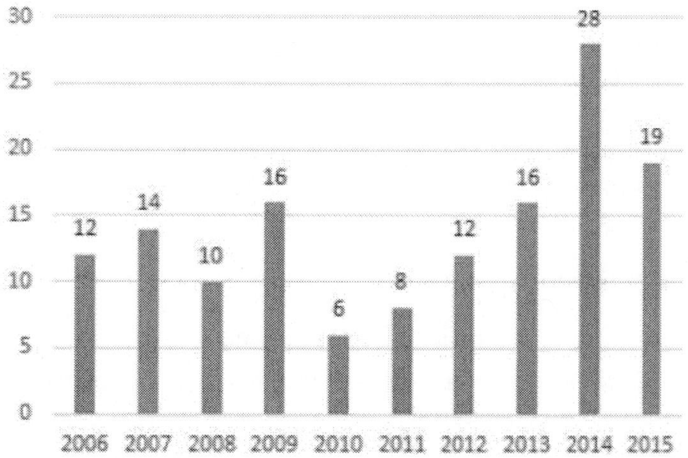

Fig. 2. Distribution of journal articles by year of publication.

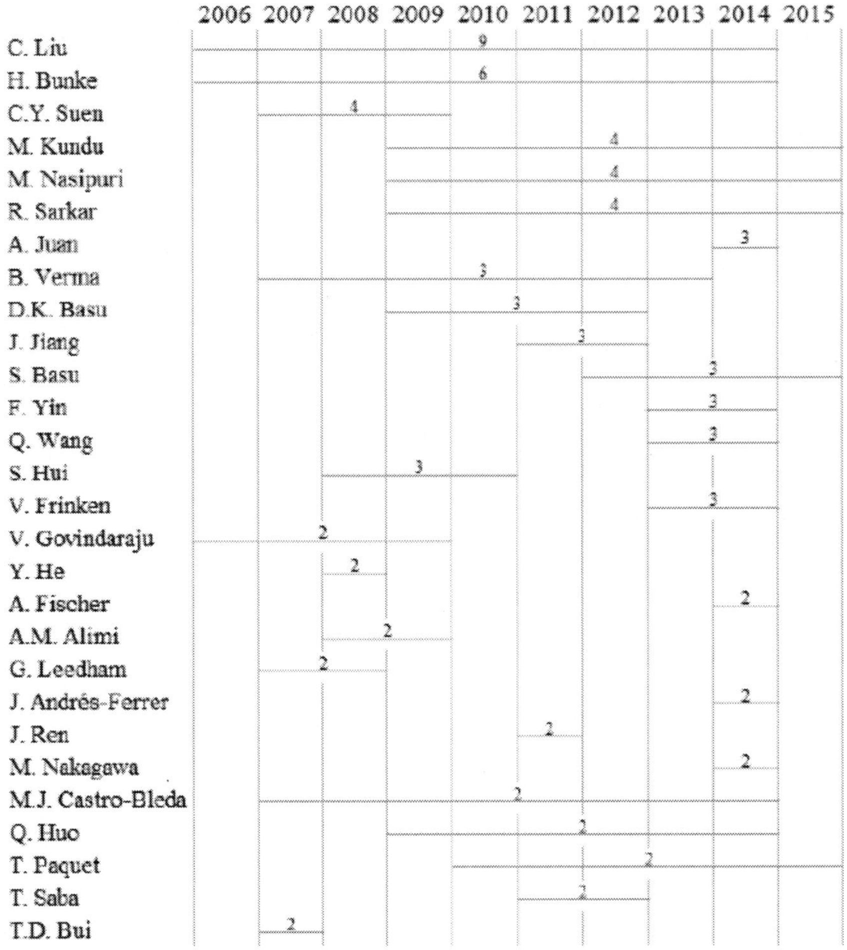

Fig. 3. Frequent author activity chart.

5.1. Distribution by Year

Distribution of the journal articles by year of publication, between 2006 and 2015, is shown in Figure **2**. There is an apparent decrease between 2009 and 2010, which is followed by a clear increase in publications between 2010 and 2014.

5.2. Distribution by Author

Figure **3** shows the frequency in which researchers authored articles. This chart can be thought of as a way to see authors' dedication to the domain and to see how consistently they report on their research.

There are 230+ unique authors among the articles we reviewed. Note that authors who only authored one paper are not shown in Figure **3**. Liu and Bunke are the most prominent authors with 9 and 6 articles respectively. Also, there are approximately 20 authors who produced 2 or 3 articles.

Years of authorship can be viewed in Figure **3** denoted by a horizontal line stretching from first to last article published. For example, Liu authored 9 articles between 2006 and 2015. In general, authors who wrote 2 or 3 papers had fewer years' spread in between with the exception of Castro-Bleda, for example. Some authors such as, Kundu, Nasipuri, and Sarkar worked together on 4 articles over the same timespan. This is shown as 3 lines spanning from the start of 2009 to the end of 2015.

Table 1. Distribution of articles by journal.

Journal Title	# Articles	Percentage %	Impact Factor
Pattern Recognition	50	35.46	2.584
Pattern Recognition Letters	26	18.44	1.062
IEEE Transactions on Pattern Analysis and Machine Intelligence	6	4.26	5.694
Expert Systems with Applications	5	3.55	1.965
Procedia Computer Science	5	3.55	-
Neurocomputing	4	2.84	2.005
Applied Soft Computing	3	2.13	2.679
Engineering Applications of Artificial Intelligence	3	2.13	1.962
AASRI Procedia	2	1.42	-
Computer and Electrical Engineering	2	1.42	0.992
Data in Brief	2	1.42	-
Image and Vision Computing	2	1.42	1.581
Information Science	2	1.42	3.893
International Journal of Advances in Engineering & Technology	2	1.42	-
Procedia Engineering	2	1.42	-

Procedia Technology	2	1.42	-
Signal Processing	2	1.42	2.238
Acta Automatica Sinica	1	0.71	-
Advanced Robotics	1	0.71	0.562
Applied Computing and Informatics	1	0.71	-
Computer	1	0.71	1.438
Digital Signal Processing	1	0.71	1.495
IEEE Sensors Journal	1	0.71	1.852
I-Manager's Journal on Software Engineering	1	0.71	-
Information Management and Computer Security	1	0.71	-
Information Processing and Management	1	0.71	1.069
Interactive Computing	1	0.71	-
Journal of System Architecture	1	0.71	0.724
Journal of the Franklin Institute	1	0.71	2.26
Knowledge-Based Systems	1	0.71	3.058
Neural Network World	1	0.71	0.412
Neural Networks	1	0.71	2.076
Optik - International Journal for Light and Electron Optics	1	0.71	0.769
Pattern Recognition and Image Analysis	1	0.71	-
Personal and Ubiquitous Computing	1	0.71	1.616
PLoS One	1	0.71	3.534

The Artificial Intelligence Review	1	0.71	0.895
The Journal of China Universities of Posts and Telecommunications	1	0.71	-
Total	141	100	

5.3. Distribution by Journal

Research papers were selected from a total of 38 different journals. Table **1** shows the distribution of articles by journal including the journal's impact factor [14]. The highest ranked journal, by impact factor, in the table is *IEEE Transactions on Pattern Analysis and Machine Intelligence* with the lowest ranked journal being *Neural Network World.*

There is a very uneven distribution of articles on HWR in the past 10 years: over 53% of the articles are published in the two journals, *Pattern Recognition* and *Pattern Recognition Letters*.

5.4. Online *vs.* Offline Techniques

There is an important distinction of handwriting recognition technique: online vs. offline. Online handwriting typically involves the use of a stylus/pen and an electronic input device, such as a tablet computer. Some other papers we encountered involve inertial measurement devices in which the user wears/holds something on/in their hand and information is transmitted back to a receiver. Online handwriting captures a time element, meaning there is order to the penstrokes. This makes it possible to do additional calculations and algorithms for a more successful recognition rate.

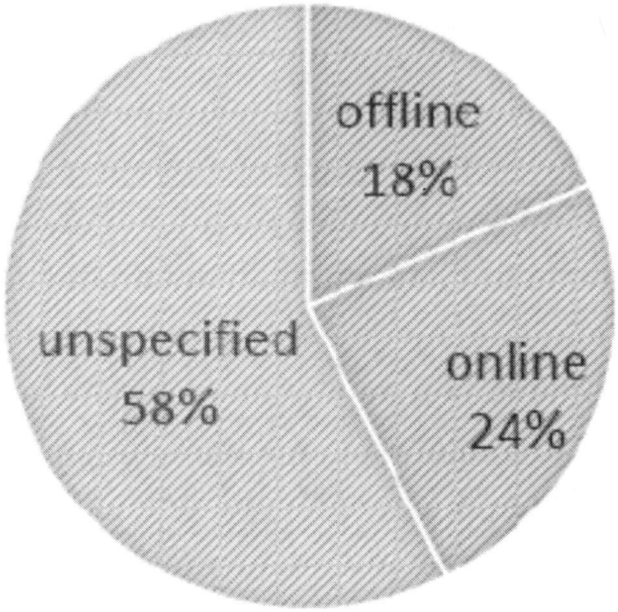

Fig. 4. Percentage of articles that use Online *vs*. Offline HWR techniques.

In contrast, offline recognition solely involves processing of image files. The image of a document can be segmented into lines, words, and sometimes characters. This method is much more challenging because the time component is omitted.

Figure **4** shows the percentage of articles that use online *vs*. offline techniques. The unspecified grouping may include articles that are about both online and offline techniques, or when a method was not mentioned. For those where the method was not mentioned, typically the article was about offline techniques. It is worth noting that offline techniques are applicable to online data images, however online techniques are not applicable to offline data.

5.5. Scripts/Languages Utilized in HWR Research

Table **2** shows the scripts upon which HWR was applied. There is some correlation, to the dataset utilized which is explained in Section 5.6. Many of the articles falling into the unspecified category in Table **2** are numerals only, English, or are not specifically applied to any script/language. Our research found that Arabic is currently the most prominent language studied for handwriting recognition, with Chinese being second. A listing of the world's most spoken languages are shown in [2], and used to verify the results achieved by this survey. Mandarin, Spanish, and English are at the top 3 of the list with Hindi/Urdu and Arabic being 4th and 5th.

Table 2. Script or Language that the article focuses on for HWR.

Script/Language	# Articles
unspecified	72
Arabic	24
Chinese or Mandarin	18
Bangla	9
English	6
Latin	4
Persian	4
Farsi	3
Sindhi	3
Amazigh	2
Malayalam	2
Devanagari	2
Amharic	1
Greek	1
Hindi	1
Japanese	1
Marathi	1
Naskh	1
Nasta'liq	1
Pashto	1
Pitman shorthand	1
Pushto	1
Renqun	1
Tamil	1
Telugu	1
Thai	1
Tibetan	1
Urdo	1

5.6. Distribution by Testing/Training Dataset

Presently, there are many datasets used for training and testing in HWR. However, many researchers have created their own datasets for their research purposes. Typically, they provided details about the number of unique writers, whether they recorded individual characters or whole words, and how big of a dataset they produced.

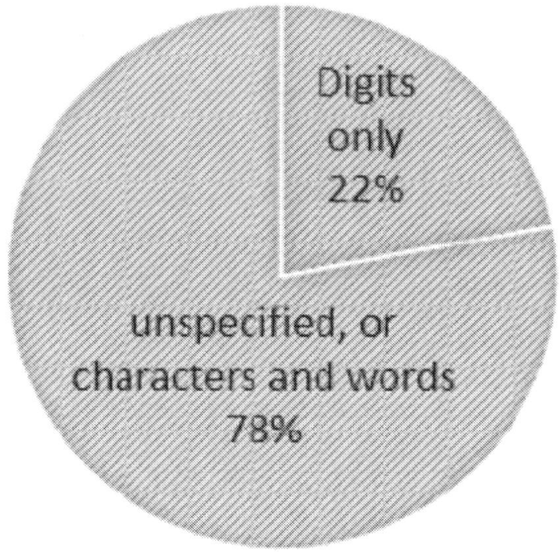

Fig. 5. Percentage of papers where techniques were applied to digits/numerals only.

Table **3** shows datasets that appear in more than 1 article. Some of these datasets are freely available for download online, while others can be obtained for a small fee. MNIST [26] is the most widely used dataset--it is a handwritten digit database. Another popular digit database is published by the United States Postal Service (USPS) and was gathered by scanning zip codes on mail envelopes. The Center of Excellence for Document Analysis and Recognition (CEDAR) dataset [45] has segmented digits and zip codes and appeared often in our review. Figure **5** shows that a very large percentage--nearly one quarter--of the articles had techniques applied to digits only.

In addition to handwritten digit datasets, word and sentence datasets are available. CASIA-HWDB and CASIA-OLHWDB [11] are Chinese datasets and can be used for research purposes only. IAM-DB and IAM-OnDB [30] are prominent English datasets consisting of words and sentences. There were six articles written during this time period that are about transcribing mathematical equations, two of which use the MathBrush dataset [24]. Additionally, three papers were written specifically to introduce new datasets to the community.

Table 3. Datasets utilized in study.

Dataset	# Articles
unspecified	33
authors created for research	22
MNIST	19
CASIA-HWDB	9
ENIT	8
IAM-DB	8
IFN	8
CEDAR	7
CMATERdb	7
USPS	7
CENPARMI	5
CASIA-OLHWDB	4
HIT-MW	3
IAM-OnDB	3
new database published	3
NIST Special Database 19	3
RIMES	3
CVPR	2
ETL-9B	2
Kuchibue	2
MathBrush	2
OLHWDB	2
UCI	2
UNIPEN	2

5.7. Distribution by Methodology

Some researchers utilize multiple (ensemble) feature extraction and classifier methods in order to achieve high accuracy when testing on the datasets. These methods are captured in Table **4**. Only methods occuring in at least 5 articles are shown in the table.

Table 4. Methodologies.

Methodology	Acronym	# Articles
Hidden Markov Model	HMM	29
Neural Network	NN	24
Support Vector Machine	SVM	19
Modified Quadrat-ic Discriminant Function	MQDF	9
Dynamic Time Warping	DTW	6
Genetic Algorithm	GA	6
Principal Compo-nent Analysis	PCA	5

The HMM method occured very frequently in our review of papers with favorable results (70% to 80% accuracy reported on several datasets). Neural Networks, also popular in our research, can be categorized into many different types. however, there was not a pattern in the types of networks. Common Neural Networks encountered include: MLPNN, TDNN, KBANN, ADFNN, FKNN, FFNN, and MTRNN, with Multi-layer Perceptron Neural Networks being the most-common.

As stated in Section 2, typically features are extracted, producing a feature vector. Next, a classifier is used to compare the distance, or similarity, between vectors. In the case of a word (rather than a single digit or character), sometimes a sliding window region of interest (ROI) approach is used to pull out narrow regions of interest. In this case, the adjacent ROI's may be analyzed by a bidirectional neural network to determine the word or solely the prior ROI's may be used in a recurrent neural network.

5.8. Special Topics

There were several special topics that appeared in the research:

- Mathematical expression [6, 8, 40, 46-48]
- Inertial [7, 18, 22]
- Signature verification [15]

Mathematical expression recognition [13] is particularly useful in an online system such as a tablet. Signature verification research may help to spawn future commercial systems that identify fraudulent attempts at payment kiosks. There are likely more articles on these special topics during the 2006-2015 time-frame, however search queries other than the ones we used may be needed to find them.

6. Conclusion

6.1. Research Implications

Several trends are apparent from our research. There has been and will likely continue to be research done on digits only. The MNIST dataset provides a good way to test digit identification methods. This is particularly useful in the commercial application of postal zip code transcription for mail sorting and bank check currency amount reading.

There is a lot of research done on Arabic scripts which is useful because there are many historical text archives. Latin script which applies to English and many other languages is also heavily studied. Chinese characters and its derivatives can be studied using datasets such as CASIA. African datasets are lacking, presenting an opportunity for researchers--for example a new dataset made available in 2015 for the Tifinagh script is published in [10].

The prominence of HMM, SVM, and Neural Networks in HWR is apparent over the past 10 years. The sliding window approach is particularly popular with HMM and NN methods. As new learning algorithms emerge, they may be applied to HWR, but for the time being these three methods will likely continue to be researched. Several papers used ensemble classifiers, meaning that they combined several classifiers and utilize a voting mechanism in order to achieve a higher accuracy rate--this will likely continue and grow.

As electronic writing on tablets becomes more popular, online techniques will become more prominent. Until then, an equal spread of online and offline techniques will likely be seen. The problem of historical document transcription is an offline and "big data" problem. Some historical documents have been meta-tagged, however searching meta-tags rather than full transcript is not as useful. Until it is solved (if it ever is), huge archives of documents which have yet to be transcribed will persist.

6.2. Future Work

Future survey work could include conference proceedings. There are many quality conference proceedings worthy of inclusion in a survey but we did not include them due to the sheer volume of papers. Furthermore, some of the conference proceedings are essentially the same as a document published in a journal.

References

1. Abuzaraida MA, Zeki AM, Zeki AM. (2010). Segmentation techniques for online Arabic handwriting Recognition: A survey. In: Information and Communication Technology for the Muslim World (ICT4M), 2010 International Conference. pp. D37–D40. DOI 10.1109/ICT4M.2010.5971900.
2. Accredited Language. (2016). The 10 most common languages. URL https://www.alsintl.com/blog/most-common-languages/, accessed: 2016-08-09.
3. Acharyya A, Rakshit S, Sarkar R, Basu S, Nasipuri M. (2013). Hand written word recognition using mlp based classifier: A holistic approach. International Journal of Computer Science, 10(2), 422-427.

4. Adankon MM, Cheriet M. (2009). Model selection for the ls-svm. application to handwriting recognition. Pattern Recognition, 42(12), 3264-3270.

5. AlKhateeb JH, Ren J, Jiang J, Al-Muhtaseb H. (2011). Offline handwrittenarabic cursive text recognition using hidden markov models and re-ranking.Pattern Recognition Letters 32(8):1081-1088.

6. Alvaro F, S´anchez JA, Bened´ı JM (2014) Recognition of on-line hand written mathematical expressions using 2d stochastic context-free grammars and hidden markov models. Pattern Recognition Letters, 35, 58-67.

7. Amma C, Georgi M, Schultz T. (2014). Air writing: a wearable handwriting recognition system. Personal and Ubiquitous Computing, 18(1), 191-203.

8. Awal AM, Mouch`ere H, Viard-Gaudin C. (2014). A global learning approach for an online handwritten mathematical expression recognition system. Pattern Recognition Letters, 35, 68-77.

9. Bahlmann C, Haasdonk B, Burkhardt H. (2002). Online handwriting recognition with support vector machines-a kernel approach. In: Frontiers in handwriting recognition. 2002. Proceedings. 8th International Workshop on, IEEE. pp. 49-54.

10. Bencharef O, Chihab Y, Mousaid N, Oujaoura M. (2015). Data set for tifinagh handwriting character recognition. Data in Brief, 4, 11-13.

11. Liu C-L, Yin F, Wang DH, Wang QF. (2011). Casia online and offline Chinese handwriting databases. In: Proc. 11th International Conference on Document Analysis and Recognition (ICDAR). pp. 37-41.

12. Casey RG, Lecolinet E. (1996). A survey of methods and strategies in character segmentation. IEEE Transactions on Pattern Analysis and Machine Intelligence, 18(7), 690-706.

13. Chan KF, Yeung DY. (2000). Mathematical expression recognition: a survey. International Journal on Document Analysis and Recognition, 3(1), 3-15.

14. Citefactor. (2016). Citefactor.org. URL https://www.citefactor.org, accessed: 2016-08-10.

15. Clarke N, Mekala A. (2007). The application of signature recognition to transparent handwriting verification for mobile devices. Information Management & Computer Security, 15(3), 214-225.

16. Fischer A, Baechler M, Garz A, Liwicki M, Ingold R. (2014). A combined system for text line extraction and handwriting recognition in historical documents. In: Proceedings - 11th IAPR International Workshop on Document Analysis Systems, DAS. pp. 71-75.

17. Frinken V, Fischer A, Mart´ınez-Hinarejos CD. (2013). Handwriting recognition in historical documents using very large vocabularies. In: Proceedings of the 2nd International Workshop on Historical Document Imaging and Processing - HIP '13, ACM Press, New York, New York, USA. pp. 67-72.

18. Hsu YL, Chu CL, Tsai YJ, Wang JS. (2015). An inertial pen with dynamic time warping recognizer for handwriting and gesture recognition. IEEE Sensors Journal, 15(1), 154-163.

19. Jayadevan R, Kolhe SR, Patil PM, Pal U. (2012). Automatic processing of handwritten bank cheque images: A survey. International Journal on Document Analysis and Recognition, 15(4), 267-296.

20. Santosh KC, Nattee C. (2009). A comprehensive survey on on-line handwriting recognition technology and its real application to the Nepalese natural handwriting. Kathmandu University Journal of Science, Engineering, and Technology, 5(I), 31-55.

21. Kessentini Y, Paquet T, Hamadou AB. (2010). Off-line handwritten word recognition using multi-stream hidden markov models. Pattern Recognition Letters, 31(1), 60-70.

22. Kim DW, Lee J, Lim H, Seo J, Kang BY. (2014) Efficient dynamic time warping for 3d handwriting recognition using gyroscope equipped smartphones. Expert Systems with Applications, 41(11), 5180-5189.

23. Koerich AL, Sabourin R, Suen CY. (2003). Large vocabulary off-line handwriting recognition: A survey. Pattern Analysis & Applications, 6(2), 97-121.

24. Labahn G, Lank E, MacLean S, Marzouk M, Tausky D. (2008). Mathbrush: A system for doing math on pen-based devices. In: Proceedings of the 2008 The Eighth IAPR International Workshop on Document Analysis Systems, IEEE Computer Society, Washington, DC, USA, DAS '08. pp. 599-606. DOI 10.1109/DAS.2008.21, URL http://dx.doi.org/10.1109/DAS.2008.21.

25. Lauer F, Suen CY, Bloch G. (2007). A trainable feature extractor for handwritten digit recognition. Pattern Recognition, 40(6), 1816-1824.

26. LeCun Y, Cortes C. (2010). MNIST handwritten digit database URL http://yann.lecun.com/exdb/mnist.

27. Lorigo LM, Govindaraju V. (2006). Offline Arabic handwriting recognition: a survey. IEEE Transactions on Pattern Analysis and Machine Intelligence, 28(5), 712-724.

28. Lu Y, Shridhar M. (1996). Character segmentation in handwritten wordsan overview. Pattern recognition, 29(1), 77-96.

29. Madhvanath S, Govindaraju V. (2001). The role of holistic paradigms in handwritten word recognition. IEEE Transactions on Pattern Analysis and Machine Intelligence, 23(2), 149-164.

30. Marti UV, Bunke H. (2003). The IAM-database: An English sentence database for offline handwriting recognition. International Journal on Document Analysis and Recognition, 5(1), 39-46.

31. Mirvaziri H, Javidi MM, Mansouri N. (2009). Handwriting recognition algorithm in different languages: survey. In: International Visual Informatics Conference, Springer. pp. 487-497.

32. Moubtahij HE, Halli A, Satori K. (2014). Review of feature extraction techniques for offline handwriting arabic text recognition. International Journal of Advances in Engineering & Technology, 7(1), 50.

33. Pal U, Jayadevan R, Sharma N, (2012), Handwriting recognition in indian re-

gional scripts: A survey of offline techniques. ACM Transactions on Asian Language Information Processing (TALIP), 11(1), 1-35,

34. Park DH, Kim HK, Choi IY, Kim JK. (2012). A literature review and classification of recommender systems research. Expert Systems with Applications 39(11), 10,059-10,072.
35. Plamondon R, Srihari SN. (2000). On-line and off-line handwriting recognition: A comprehensive survey. IEEE Transactions on Pattern Analysis and Machine Intelligence, 22(1), 63.
36. Pl¨otz T, Fink GA. (2009). Markov models for offline handwriting recognition: a survey. International Journal on Document Analysis and Recognition (IJDAR), 12(4), 269-298.
37. Rodr´ıguez-Serrano JA, Perronnin F. (2009). Handwritten word-spotting using hidden markov models and universal vocabularies. Pattern Recognition, 42(9), 2106-2116.
38. Sadri J, Suen CY, Bui TD. (2003). Application of support vector machines for recognition of handwritten arabic/persian digits. In: Proceedings of Second Iranian Conference on Machine Vision and Image Processing, vol. 1, pp. 300-307.
39. Shanjana C, James A. (2015). Offline recognition of malayalam handwritten text. Procedia Technology, 19, 772-779.
40. Simistira F, Katsouros V, Carayannis G. (2015). Recognition of online handwritten mathematical formulas using probabilistic svms and stochastic context free grammars. Pattern Recognition Letters, 53, 85-92.
41. Srihari SN. (1993). Postal processing and character recognition recognition of handwritten and machine-printed text for postal address interpretation. Pattern Recognition Letters, 14(4), 291-302.
42. Tagougui N, Kherallah M, Alimi AM. (2013). Online arabic handwriting recognition: A survey. International Journal on Document Analysis and Recognition (IJDAR), 16(3), 209-226.
43. Tappert C, Suen C, Wakahara T. (1990). The state of the art in online handwriting recognition. IEEE Transactions on Pattern Analysis and Machine Intelligence, 12(8), 787-808.
44. Tappert CC, Suen CY, Wakahara T. (1988). Online handwriting recognitiona survey. In: 1988 Proceedings 9th International Conference on Pattern Recognition. pp. 1123-1132.
45. University of Buffalo. (2016). Center of excellence for document analysis and recognition (cedar). URL http://www.cedar.buffalo.edu/Databases/, accessed: 2016-10-07.
46. Vuong BQ, Hui SC, He Y. (2008). Erratum to progressive structural analysis for dynamic recognition of on-line handwritten mathematical expressions [Pattern Recognition Letter, 29(5), 647-655]. Pattern Recognition Letters, 29(9), 1454.
47. Vuong BQ, Hui SC, He Y. (2008). Progressive structural analysis for dynamic recognition of on-line handwritten mathematical expressions. Pattern Recogni-

tion Letters, 29(5), 647-655.

48. Vuong BQ, He Y, Hui SC. (2010). Towards a web-based progressive handwriting recognition environment for mathematical problem solving. Expert Systems with Applications, 37(1), 886-893.

Verification of Hypotheses Generated by Case-Based Reasoning Object Matching

Petra Perner

Institute of Computer Vision and Applied Computer Sciences, IBaI
PSF 301114, 04251 Leipzig, Germany
pperner@ibai-institut.de, www.ibai-institut.de

Abstract. Case-based reasoning object-matching is the methods at choice when the objects can be identified by case models. The result of the matching process will be a number of hypothesis for the true shape of the objects. These hypothesizes have to be verified in a hypothesis-verification process. We review in this paper what has been done so far and present our hypothesis-verification rules. The rules get evaluated and the results are shown in the images and the achievements are discussed. We consider two different hypothesis-verification rules, one is based on set-theory and the other one is based on statistical measures. Finally, we describe the results achieved so far and give an outlook about further work.

Keywords: Case-based reasoning object-matching, hypothesis-test verification, set theory, statistical measures

1. Introduction

Case-based object-matching [1] is the methods at choice when the objects can be identified by case models. These case models can be learnt from the raw data by case mining [2]. For the case-matching procedure, we need a proper similarity measure that depends of the case model description. In our case, the case models are object contours such as round, ellipse-like, or more fuzzy-like geometric figures. The chosen similarity measure in this work is the cosine-similarity measure [3]. The properties of this similarity measure have been described in detail in [3]. The case matcher takes the case models and matches them against the objects in the image. In case the similarity measure is high the found contour will be marked in the image. Often the matcher does not bring out only one contour for an object, instead of the matcher fires several times at slightly different spatial positions in the image for the same object. These multiple matches have to be evaluated after the matching in a hypothesis verification procedure. The aim of this hypothesis-verification procedure is to obtain only the considered object.

We describe in this paper what kind of hypothesis verification methods we have developed and tested on our image database. The state-of-the-art of hypothesis verification methods is described in Section 2. In Section 3 we describe the hypothesis generation and the problems concerned with it. In Section 4 we describe kinds of hypothesis-verification based on Set Theory. We give results in Section 5. Hypothesis-verification rules based on statistics and results are given in Section 6. Finally we summarize our work in Section 7 and we give an outlook to further work on improvements of the matching results.

2. State-of-the Art

The aim of the hypothesis verification is to decide whether a match can be accepted as correct or not. Therefore hypothesis verification is closely related to the object recognition process. In literature we can find different approaches for this process. Grimson and Huttenlocher [1] as well as Jurie [2] and Kartatzis *et al.* [3] refer to features in the form of points or line segments. The common target of all these researchers is to find the best pose for the detected data features. However, all papers follow different strategies: Grimson and Huttenlocher [1] focus on the question how random matches can be prevented. They developed a formal means for finding the fraction of model features that have to be evaluated in order to ensure that the match occurs only with a given probability at random. The derivation of this fraction is done in three steps whereas the type of feature, the type of transformation from model to image and a bound on the positional and orientational error are known. First for every pairing of a model feature to data feature the set of transformations is determined. This set defines a particular volume in the transformation space. In the next step the probability of a common point of intersection between l and more volumes is calculated. This probability corresponds to match of at least l pairings of model and image features. Last a second probability that describes that l or more volumes will intersect at random, is used to specify a threshold for the fraction of model features that have to be evaluated at least in order to ensure that the probability of a random match is lower than a given value.

The aim of the research of Jurie [2] is to find the pose of the model features that best matches the data features. The pose hypotheses are generated by correspondences between the model and the data features. Early researches propose to evaluate only some correspondences in order to find an initial pose hypothesis P that is refined by iteratively enlarging the number of correspondences. Jurie [2] describes that this way of hypothesis generation and verification is not optimal. Therefore the paper suggests the opposite approach: A pose space is generated from different model-data-pairings. A "box" of the pose space is computed including the initial position P that is large enough to compensate the data errors. Assuming that the distribution of model-data-correspondences is Gaussian, the maximal probability of the object to be matched is determined. Then the box can be refined. The process repeats until the "box" only contains one pose.

A simpler method of model-based pose estimation and verification is described in Shahrokni *et al.* [4]. They deal with the automatic detection of polyhedral objects. Hypotheses are generated by the knowledge-based connection of corners and line segments. The model and the transformed hypotheses are evaluated with the method of the least squares. The best hypothesis minimizes the sum of squared differences between the model and the transformed hypothesis.

Katartzis *et al.* [3] discuss the automatic recognition of rooftops, which are characterized by lines and their connections. After detecting the line segments in the image, they are grouped in a hierarchical graph. The highest level of the hierarchy contains closed contours. Every node of the graph is assigned a value that on the one hand assigns the saliency of the hypothesis and on the other hand represents the likelihood of the presence of a 3D-structure, which depends on domain-specific knowledge. Based on the hierarchical graph is defined a Markov Random Field (MRF). By maximization of the a posteriori probability of the MRF for the concrete graph a consistent configuration of the data is found.

In general the verification process for object hypotheses based on line segments is a widely discussed field.

An approach that totally differs from the discussed ones is given in Leibe *et al.* [7]. The heart of the described object recognition system is a database with different appearances of parts of the object that should be recognized. Additionally an "Implicit Shape Model" is learnt in order to combine the parts to a correct object. If multiple objects are located in the image then some hypotheses may overlap each other so that a verification step is required. The method follows the principle of Minimal Description Length (MDL) that is borrowed from the information theory. The description length of a hypothesis depends on its area and the probability that the pixels inside the hypothesis are no object pixels. The description length of two overlapping hypotheses is generated in the same way. From the resulting values is concluded whether two overlapping hypotheses refer to two objects or only to one.

3. Hypothesis Generation and Problems

3.1. Hypothesis Generation

In this Section, we want to give you an overview about the model-based object recognition method that we use to generate our hypotheses. The method is extensively discussed in Perner and Buehring [8].

A model-based object recognition method uses templates that generalize the original objects and matches these templates against the objects in the image. During the match a score is calculated that describes the goodness of the fit between the object and the template.

We determine the similarity measure based on the cross correlation by using the direction vectors of an image. This requires the calculation of the dot product $\widetilde{l}_k$ between each direction vector of the model $\vec{m}_k = (v_k, w_k)^T$, $k = 1, \ldots, n$, and the corresponding image vector $\vec{i}_k = (d_k, e_k)^T$:

$$\widetilde{l}_k = \langle \vec{m}_k, \vec{i}_k \rangle = \vec{m}_k \cdot \vec{i}_k = (v_k \cdot d_k + w_k \cdot e_k), \quad k = 1, \ldots, n \qquad (1)$$

Note that the dot product $\widetilde{l}_k$ (see Equation 1) takes also into account the length of the vectors $\vec{m}_k$ and $\vec{i}_k$. That means that $\widetilde{l}_k$ is influenced by the intensity of the contrast in the image and the model. In order to remove this influence, the direction vectors are normalized to unit length by dividing them through their gradient:

$$l_k = \left\langle \frac{\vec{m}_k}{\|\vec{m}_k\|}, \frac{\vec{i}_k}{\|\vec{i}_k\|} \right\rangle = \frac{\vec{m}_k \cdot \vec{i}_k}{\|\vec{m}_k\| \cdot \|\vec{i}_k\|} = \frac{v_k \cdot d_k + w_k \cdot e_k}{\sqrt{v_k^2 + w_k^2} \cdot \sqrt{d_{k_i}^2 + e_k^2}}, \quad k = 1, \ldots, n \qquad (2)$$

The score l_k (see Equation 2) takes into account only the directions of the model and the image vector, *i.e.* it is invariant against changes of the contrast intensity. We can get the angle between the direction vectors by determining the value of l_k. Therefore we can conclude that the value of l_k ranges from -1 to 1. The vectors $\vec{m}_k$ and $\vec{i}_k$ have the same direction if $l_k = 1$, the vectors are orthogonal if $l_k = 0$ and both vectors have opposite direction if $l_k = -1$. In the rest of the paper we say that the value of l_k is the *local similarity score* of the two direction vectors $\vec{m}_k$ and $\vec{i}_k$.

Usually, we are mainly interested in the similarity score between the complete model and the image. We want to define this *global similarity score* s_1 between the model and the image as the mean of all local similarity scores:

$$s_1 = \frac{1}{n} \sum_{k=1}^{n} l_k \qquad (3)$$

Just like the local similarity score l_k the global score s_1 is invariant against illumination changes and it ranges from -1 to 1. In case of $s_1 = 1$ and $s_1 = -1$ the model and the image object are identical. If $s_1 = 1$ then all vectors in the model and the corresponding image vectors have exactly the same direction. If $s_1 = -1$ then all the vectors have exactly opposite directions, that is only the contrast between the model and the image is changed.

In general we have to subdivide between global and local contrast changes. If the contrast between the model and the image is globally inversed then all the model and image vectors have opposite directions. If the contrast is locally inversed then only some model and image vectors have opposite direction. With some little modifications the similarity measure s_1 becomes invariant to global contrast changes (see Equation 4) and local contrast changes (see Equation 5), respectively.

$$s_2 = \left| \frac{1}{n} \sum_{k=1}^{n} l_k \right| \qquad (4)$$

$$s_3 = \frac{1}{n} \sum_{k=1}^{n} |l_k| \qquad (5)$$

In contrast to range of s_1, the values of s_2 and s_3 are non-negative.

The aim is to store only one model for objects with similar shapes of different scale and rotation. Therefore a transformed model must be compared to the image at a particular location. The value of $\arccos s_2$ indicates the mean angle between the model and the image vectors.

3.2. Kinds of Hypothesis Verification Based on Set-Theory

The matching process determines each possible match between the image pixels and the model. In the following we consider the found object as a hypothesis. To each hypothesis is assigned a matching score based on the similarity measure s_3. The score

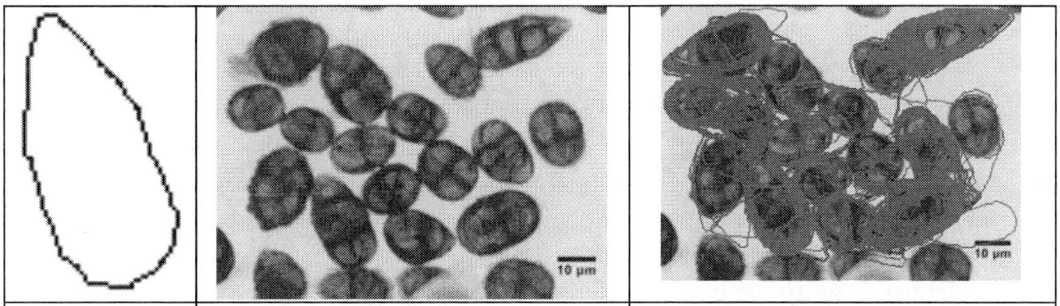

a. Model | b. Original Image | c. Hypothesized Objects

Figure 1. Contour model, original image and hypothesized objects.

Table 1. Relationships between two Hypotheses.

Relation	Figure	Description
a. Inside	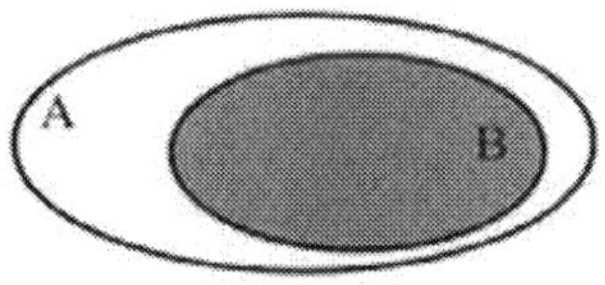	We say that the set $S(B)$ is ***inside*** the set $S(A)$ if all elements that are included in the set $S(B)$ are also included in the set $S(A)$, i.e. $S(B) \subset S(A)$ and there is $S(A) \cap S(B) = B$ and $S(A) \cup S(B) = S(A)$.
b. Overlapping	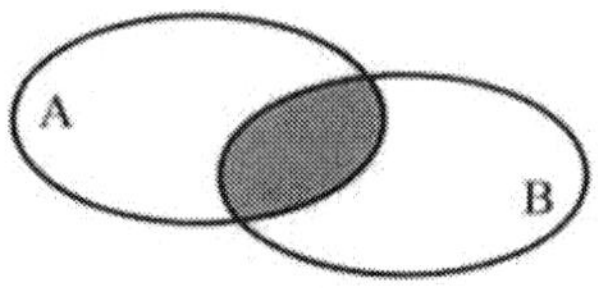	We say that the sets $S(A)$and $S(B)$ ***overlap*** each other if they have some equal elements, i.e. $S(A) \cap S(B) \neq \emptyset$, $S(A) \cap S(B) \neq S(A)$ and $S(A) \cap S(B) \neq S(B)$.
c. Almost Inside	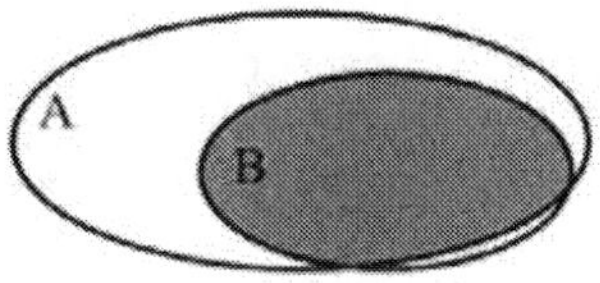	We say that the hypothesis B is ***almost inside*** the hypothesis A if almost all elements of $S(B)$ are also elements of $S(A)$, i.e. $S(A) \cap S(B) \neq \emptyset$, $S(A) \cap S(B) \neq S(A)$ and $S(A) \cap S(B) \neq S(B)$ and $\lvert S(A) \cap S(B) \rvert \approx \lvert S(B) \rvert$ and $\lvert S(A) \cup S(B) \rvert \approx \lvert S(A) \rvert$. This relation is a special case of the relation "overlapping".

d. Touching	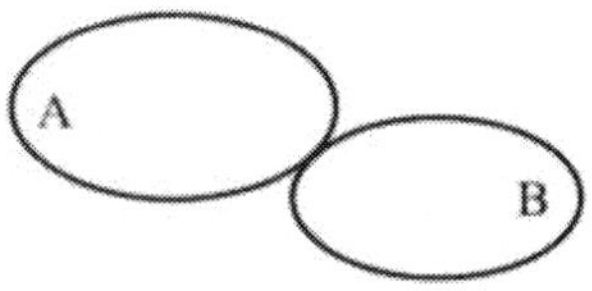	We say that hypotheses *A* and *B* are ***touching*** if their contours *C*(*A*) and *C*(*B*) have some equal elements, whereas the equal elements are neighboured. Touching Hypotheses are a special case of overlapping hypotheses. But it is also possible that two sets are touched and one set is inside the other set.

From the observation in Figure **1** we can see that the models often match the same object, *i.e.* we have a superimposition of models. All the hypotheses in this image have scores greater than *0.8*. Now we need to find a rule which allows us to remove determines the similarity between the model and the image pixels. It can range from *0* to 1 whereas the value of 1 says identity and the value of 0 dissimilarity. By defining a threshold for the score we can exclude hypotheses. This is the simplest hypothesis verification process. If the threshold is set to 0.8 then 734 hypotheses remain. They are shown in Figure **1** false hypotheses. The hypotheses in this particular case overlap, touch or are inside of each other. From that we can develop special relationships of the hypotheses.

The definition of the relationships is based on two hypothesized objects *A* and *B*. *S(A)* is the set of all image pixels that are inside the contour of the object *A* including also the image pixels of the contour. Equally, *S(B)* is the set of all image pixels inside the contour of object *B* including all image pixels of the contour. We want to distinguish between three relationships that are described in Table **1**.

Determining the number of common pixels of every pair of hypotheses we conclude to their relation. In Section 4 we analyse the generated hypotheses using the defined relationships.

4. Results

This Section focuses on the reduction of initial hypotheses.

4.1. The Relationship "Inside"

In this Section we want to investigate in the relationship "inside" (Table **1a**). Given a sorted list[1] of hypotheses, we first extract the hypothesis pairs that fulfil the relationship "inside" (see Figure **2b**). From the 734 hypothesized matches we can create 138 hypotheses pairs that fulfil the relationship inside, i.e. one of both hypotheses is totally overlaid by the other. Note that these pairs are only based on 136 different hypotheses. Thus we conclude that some hypotheses with high area overlay more than one smaller hypothesis. In other words we can say that most of the hypotheses are involved in more than one pair.

---

[1] The matching process lists all matched objects sorted by the scale whereas objects with the same scale are sorted by the rotation.

For the reduction process we rule that the hypothesis with the higher matching score remains while the other hypothesis of this pair is removed. Since the removed hypothesis often is a partner in more than one pair, for some hypotheses there will be no other correct "inside" partner. That is in practice that the number of hypotheses pairs may decrease for more than one pair. We want to illustrate this fact based on Figure **2**: We obtain a reduction of 67 hypotheses (Figure **2a** and **2d**) if we successively remove the "inside" partner with the lower score. Considering the hypotheses that are used to create correct "inside" pairs (Figure **2b**) and their remaining partners (Figure **2c**), gives a reduction by 95 hypotheses.

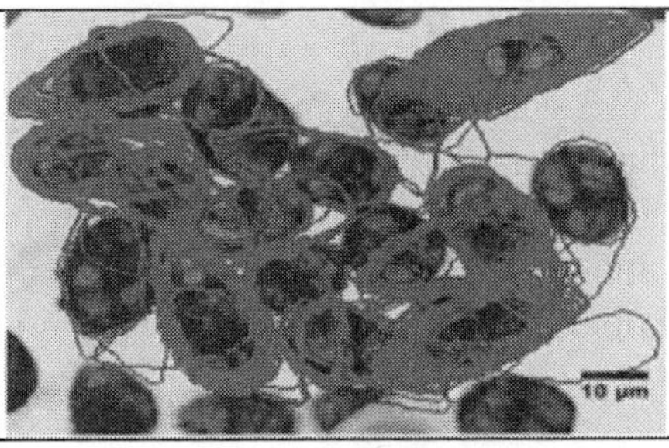

Fig. 2a. Hypothesized Matches (734 Hypotheses).

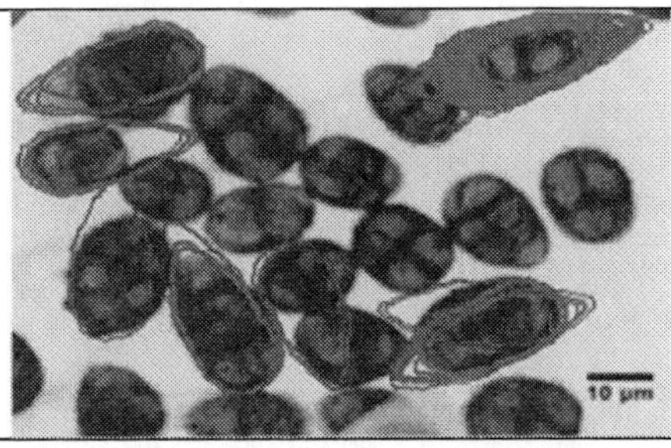

Fig. 2b. Hypothesis Pairs that fulfils the Relation "inside" (138 Pairs based on 136 Hypotheses).

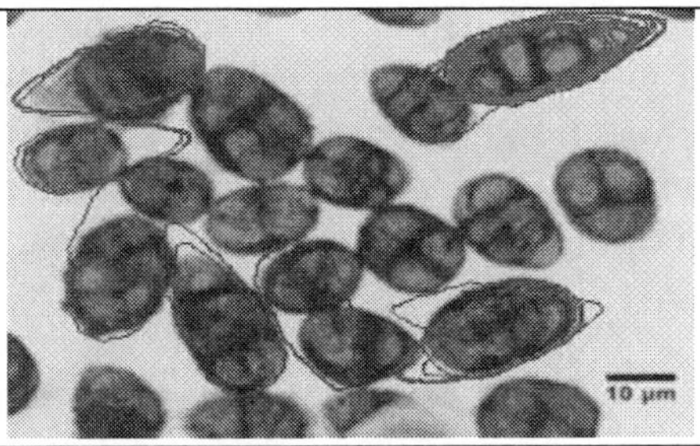

Fig. 2c. Remaining Hypotheses after removing the Hypotheses with lower Score (41 Remaining Partners).

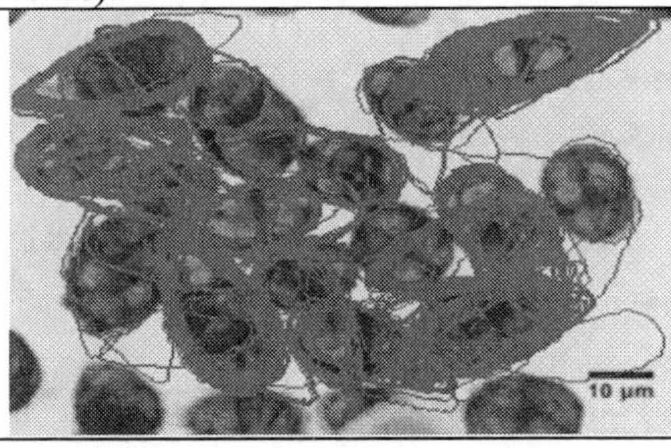

Fig. 2d. Remaining Hypotheses after applying the "Inside"- Criterion (667 Hypotheses).

From Figure **2a** and **2d** we can see that the total number of hypotheses is only reduced by about 10 %. Since this reduction does not significantly simplify the hypothesis verification process we investigate in the relation "overlapping" (see Section 3.2)

4.2. The Relationship "Overlapping"

In this Section, we concentrate on the relationship "overlapping". Although we could not significantly reduce the number of hypotheses by applying the relationship "inside" we work with the reduced number of hypotheses (see Figure **2d**). For presentation purposes we first only consider the 41 remaining hypotheses of the "inside" pairs which are shown in Figure **3** (compare to Figure **2c**). At the end of this Section we extend our investigations to the whole set of hypotheses.

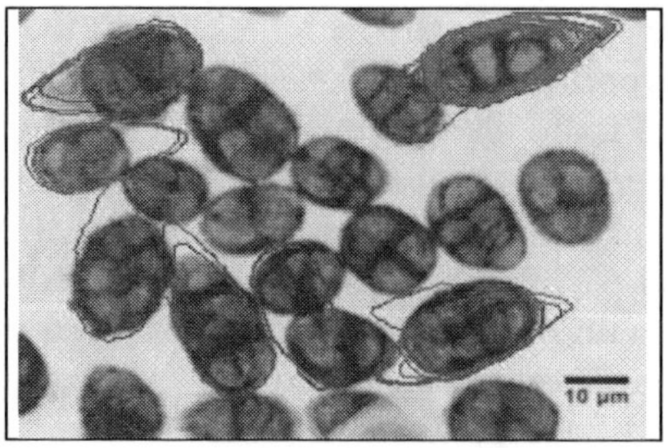

Fig. 2. Basic Hypotheses.

Note that the hypotheses are concentrated in some regions of the image (see Figure **3**). It seems that many hypotheses are slightly transformed (shifted or twisted) with respect to other hypotheses. This means that the intersection area of two overlapped hypotheses A and B has approximately the same size as the area of the hypothesis A and the hypothesis B respectively. We express this fact with the condition (6):

$$|S(A) \cap S(B)| \geq t|S(A)| \quad \text{AND} \quad |S(A) \cap S(B)| \geq t|S(B)|, \quad t \in [0,1] \qquad (6).$$

We restrict the size of the intersection area with respect to the size of the hypothesis area by the value t. If two hypotheses fulfil the Condition (6) we say they overlap each other. As well as in the discussion of the relationship inside, we assume that the best match for an object has the highest score. From two overlapping hypotheses we therefore remove the hypothesis with the lower matching score. In the first part of Figure **4** the remaining hypotheses are given when we use the overlapping condition (6) and varying the thresholds t of the minimal common hypotheses area.

Since the condition (6) mainly combines hypotheses with similar size, we replace the "AND" with "OR" (7). Then we repeat the test with the new condition (7). The results are given in the second part of Figure **4**.

$$|S(A) \cap S(B)| \geq t|S(A)| \quad \text{OR} \quad |S(A) \cap S(B)| \geq t|S(B)|, \quad t \in [0,1] \qquad (7).$$

C	t=0.95	t=0.9	t=0.8	t=0.67
(6)	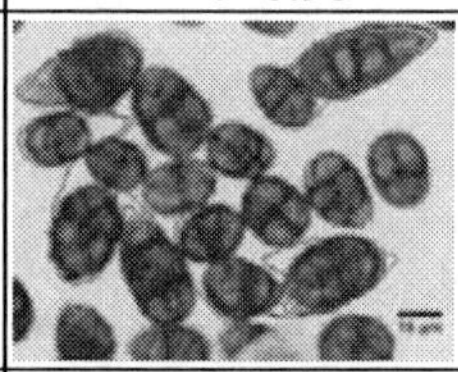	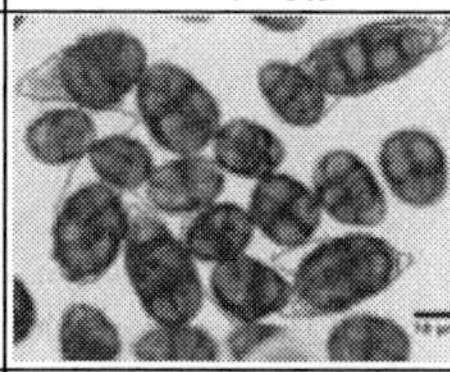	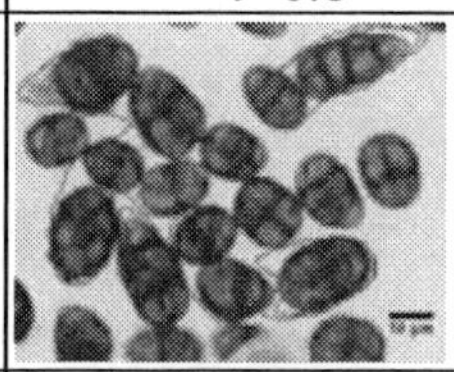	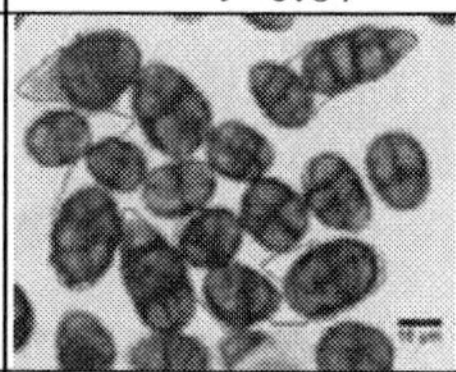
	28 Hypotheses	19 Hypotheses	14 Hypotheses	9 Hypotheses
(7)	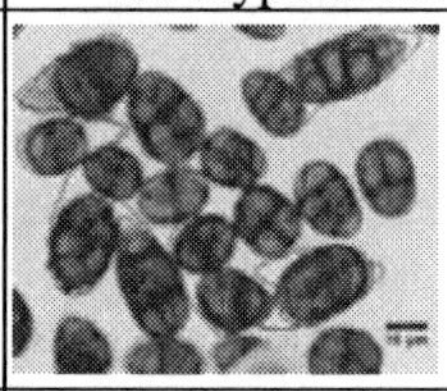	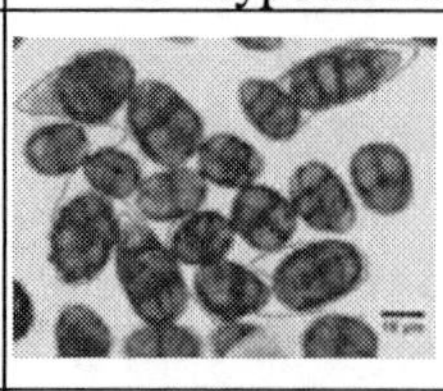	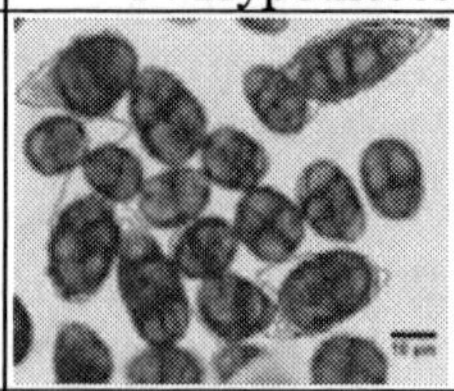	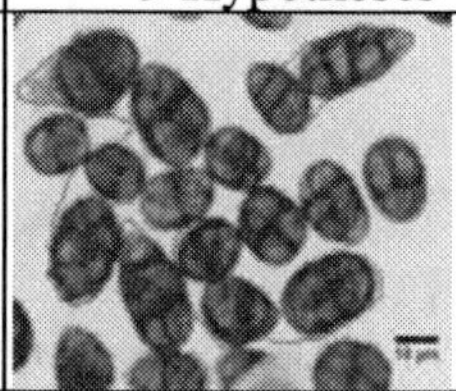
	19 Hypotheses	13 Hypotheses	9 Hypotheses	8 Hypotheses

Fig. 3. Applying the Relationship "Overlapping" with different conditions and degrees of common area to some selected Hypotheses (C = Condition).

As we can see from Figure **4** the number of hypotheses is as fewer as lower the threshold of the minimal common hypotheses area is. As we expected more hypotheses are removed with the second condition (7). Therefore we take the condition (7) using the threshold t=0.67 for applying the "overlapping" relation to all remaining hypotheses of the "inside" relation (Figure **5**).

a. Basic Hypotheses	**b.** Remaining Hypotheses
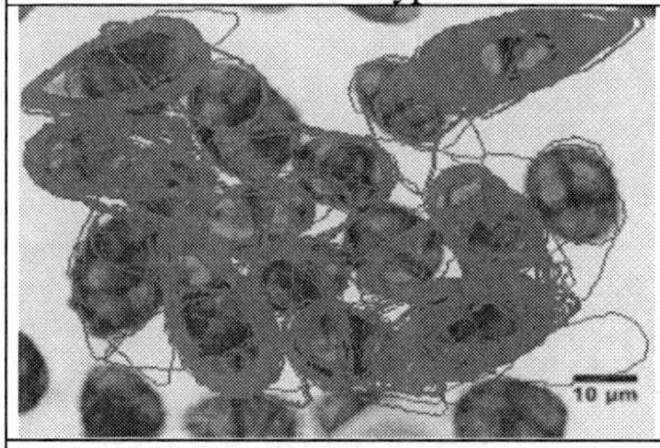	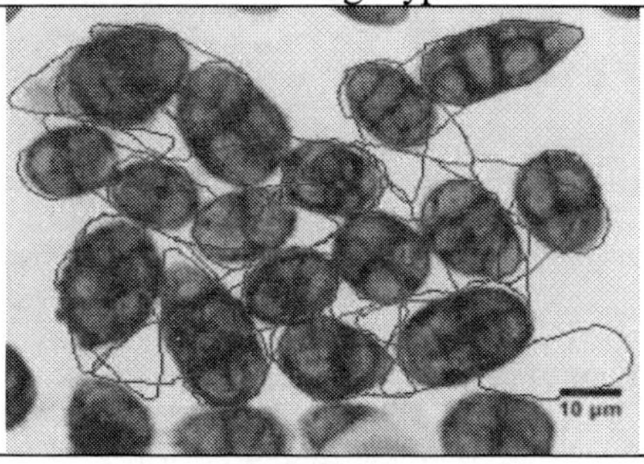
667 Hypotheses	35 Hypotheses

Fig. 4. Applying the Relationship "Overlapping" with Condition (7) and the Threshold t=0.67 after the Relationship "Inside".

From Figure **5** we can see that the applied criterion reduces the given hypotheses to about *5 %*. In order to improve the performance of the reduction process we used the described rule also to all hypothesized matches. We obtain the same result as if we remove some hypotheses with the "inside" relation. Therefore we conclude that the first step in each hypothesis verification process should be the reduction of hypotheses using "overlapping" relation defined with condition (7) and the common area threshold $t = 0.67$. From each pair of overlapped hypotheses the hypothesis with the lower matching score is removed.

5. Statistical Reduction of the Hypotheses

The method for hypothesis reduction that is described in Section 3.2 shows very good performance. One of the main weaknesses of this method is the arbitrary fixed threshold of common area. In this Section we want to discuss some more possibilities of hypothesis reduction based on statistical measures.

5.1. Common Statistical Measures

In order to determine the distribution of the matching scores that range from *0.8* to *1*, we generate a histogram by subdividing the range into non-overlapping classes with a class width of *0.05*. The number of hypotheses within each class is displayed in Figure **6**.

From the Histogram in Figure **6** we cannot conclude to any distinct distribution because we only investigate in a part of the score space. Nevertheless we can determine the mean μ_s and the standard deviation σ_s of all scores. In order to optimize the threshold we develop a criterion for removing some hypotheses based on the mean and the standard deviation (8):

$$s_i < \mu_s + f \cdot \sigma_s \Rightarrow \text{delete hypothesis } i, \quad i = 1, \ldots, h \qquad (8).$$

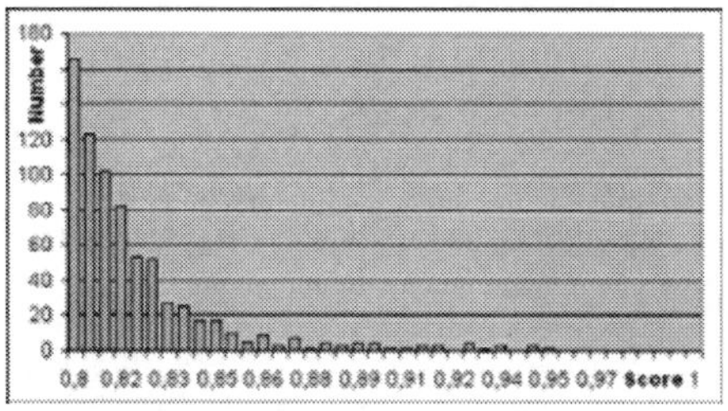

Mean = 0.8223
Standard Deviation = 0.0265

Fig. 5. Histogram, Mean and Standard Deviation of the Hypothesis Scores.

The number of hypotheses is denoted by h. Figure **11** shows the remaining hypothesized objects applying different factors f to the 734 hypotheses shown in Figure **1c**:

a. Hypotheses which Score mean > (f=0)	b. Remaining Hypotheses applying f= 1	c. Remaining Hypotheses applying f= 2	d. Remaining Hypotheses applying f= 3
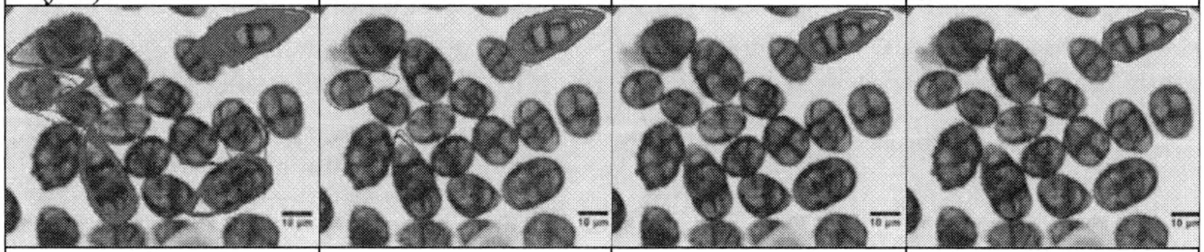			
236 Hypotheses	78 Hypotheses	39 Hypotheses	23 Hypotheses

Fig. 6. Remaining Hypotheses Applying Condition (9) with different Factors f.

From Figure **7** we can see that the number of hypotheses can be reduced significantly if the threshold for the matching score is increased. Since we expect high scores if the model matches an object very well, we will obtain as more hypotheses for one object as better the model matches the object. The results reported in Figure **7** verify this assumption.

Remember that the matcher tolerates object occlusion and touching until *20%* if the score threshold is *0.8*. The described method for hypotheses reduction increases the threshold for accepting a match as a hypothesis. In the consequence it removes also hypotheses of objects which are occluded or touched. If we want to consider also such objects, we must not reduce the hypothesized matches based on Equation (8).

5.2. Hypotheses Reduction by the Evaluation of the Local Score

For each hypothesis we can store, during the matching process, the number of model pixels that have the same local contrast as the corresponding image pixel. In the following we denote this number with c_{same}. Remember that the model and the image pixel have the same local contrast if their dot product is positive. Otherwise their local contrast is inversed. In order to measure the quality of the hypothesis we determine, with respect to the number n of model pixels, the fraction of contour pixels of the hypothesis that have the same contrast as the model. Depending on the global contrast between the model and the hypotheses we should accept very high and very low values of this fraction. Given a threshold t the minimal fraction of the same contrast for acceptance, we determine the remaining hypotheses based on Condition (9):

$$t < \frac{c_{i,same}}{n} < (1-t) \Rightarrow \text{remove hypothesis } i, \quad t \in [0,0.5], i = 1,\ldots,h \tag{9}$$

Figure shows the remaining hypotheses using different values of the threshold t.

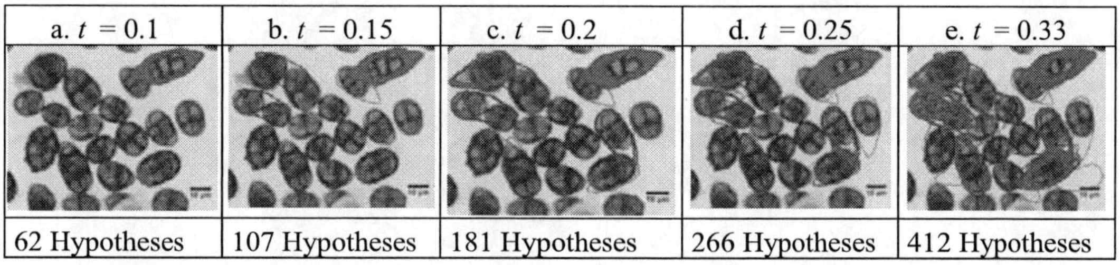

Fig. 7. Remaining Hypotheses Applying Condition (9) with different Thresholds t.

It is strange that hypotheses which seem to match the object well are earlier removed than hypotheses which include some background. Figure **8** shows this phenomenon for the hypotheses of two selected objects in the image based on the threshold $t = 0.25$. Each image in Figure **8** shows a labelled hypothesis whereas red parts display negative local contrast and blue parts positive local contrast. Below each image the relative fraction of negative local contrast is given (blue parts of the contour).

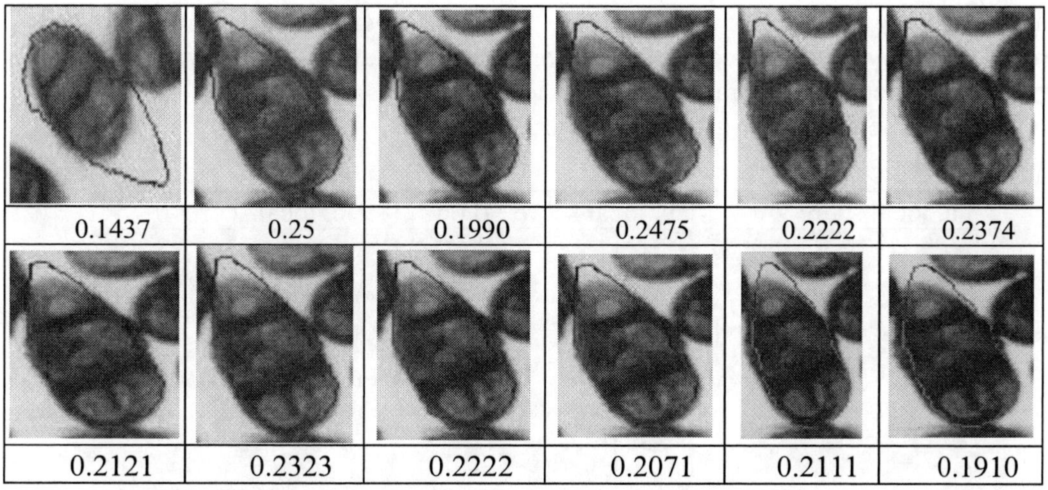

Fig. 8. Results for binary contrast changes for different Hypotheses.

Because of these unexpected results we ask if the approach of the differentiation between positive and negative local contrast is correct. The power of the similarity measure that we use is its invariance to local contrast changes, that is, we eliminate the influence of the sign of the local score by summing up the absolute amount of the local scores. On the other side we exactly evaluate the sign if we calculate the score of binary contrast changes (see Figure **9**).

Figure **10** shows the same hypotheses as Figure **9**, but now pixels which local score higher than *0.9* or lower than *-0.9* are marked blue. The other parts are red. Below the images the relative faction of pixels with high local score *(>0.9)* is given.

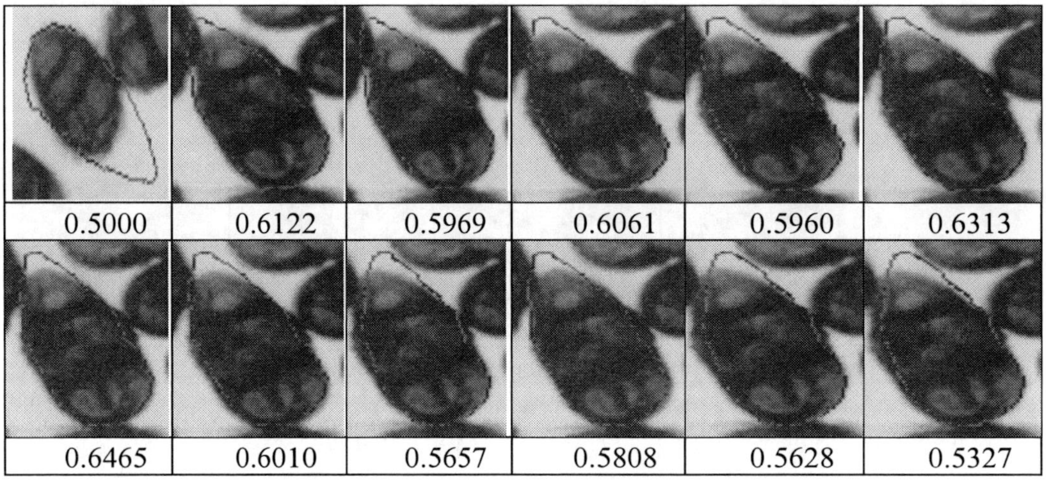

Fig. 9. Results for local score higher than 0.9 and lower than -0.9.

Note that the fraction of high local scores depends on the threshold that defines a pixel with "*high local score*". As higher as the threshold is as lower is the value of this fraction.

Similar to the experiment we carried out for the binary contour based on contrast changes (see Figure **8**) we now create the binary contour for all hypotheses as described above by thresholding the local scores at the value *0.9*. Then we remove the hypotheses which relative fraction of high local scores is lower than (a) *0.6* (b) *0.66* and (c) 0.75. The results are given in Figure **11**.

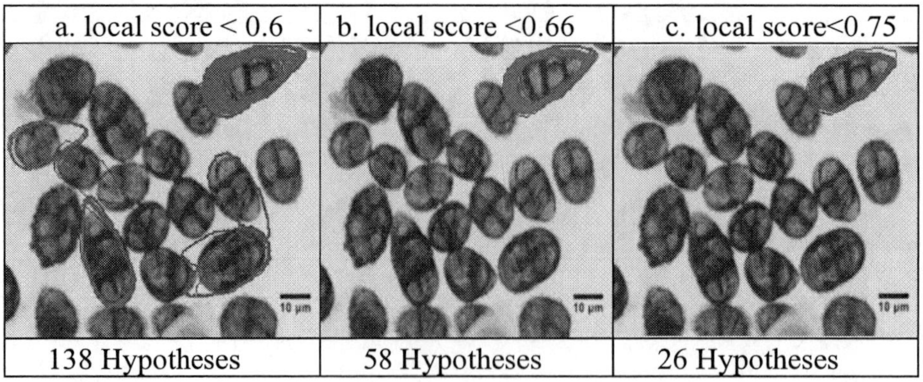

Fig. 10. Results for different local scores.

From Figure **11** we can see that this method is another possibility to reduce the number of hypotheses. Since the hypotheses around the object in the right upper corner have high similarity scores most of these hypotheses remain.

6. Conclusions

In this paper we have described our hypothesis verification process for our case-based reasoning shape-object matching procedure. We have described the hypothesis-generation process in brief and the problems concerned with it. Then we described the kinds of hypothesis verification we have developed for our matching procedure. Results are given for the different rules. Finally, we introduce some statistical measures for hypothesis reduction and give results. The final results show good performance but we can still think of some other verification measures that will further improve the results. These verification measures will be based on grouping the hypotheses, evaluation of the local similarity, and the background fraction of the found objects. This work is left for a further paper that will finish the hypothesis-verification work and will give a good summary about the work.

References

1 W.E.L. Grimson and D.P. Huttenlocher: On the Verification of Hypothesized Matches in Model-Based Recognition, IEEE Transactions on Pattern Analysis and Machine Intelligence, 13 (12), pp. 1201-1213, 1991.

2 F. Jurie: Hypothesis Verification in Model-Based Object Recognition with a Gaussian Error Model, In: Proc. of the European Conference on Computer Vision, Freiburg, Germany, pp. 642-656, 1998.

3 A. Katartzis, H. Sahli, E. Nyssen, and J. Cornelis: Detection of Buildings from a Single Airborne Image using a Markov Random Field Model, In: Proc. of IEEE International Geoscience and Remote Sensing Symposium (IGARSS 2001), Sydney, Australia, 2001.

4 A. Shahrokni, L. Vacchetti, V. Lepetit, and P. Fua: Polyhedral Object Detection and Pose Estimation for Augmented Reality Applications, 2002.

5 B. Leibe, A. Leonardis, and B. Schiele: Combined Object Categorization and Segmentation with an Implicit Shape Model, In: ECCV'04 Workshop on Statistical Learning in Computer Vision, Prague, May 2004.

6 P. Perner and A. Buehring: Case-Based Object Recognition, ECCBR 2004, Advances in Case-Based Reasoning, Volume 3155 of the series Lecture Notes in Computer Science pp 375-388

7 M. Petrou, P. Bosdogianni: Image Processing – The Fundamentals; John Wiley & Sons, Chichester (GB), 1999.

8 S. Wu and A. Amin: Automatic Thresholding of Gray-level Using Multi-stage Approach, In: Proc. of the 7th International Conference on Document Analysis and Recognition (ICDAR), 2003.

A Hazy Image Database with Amplitude Spectra Analysis

Shuhang Wang[1], Yu Tian[2], Tian Pu[3], Patrick Wang[4], Petra Perner[5]

[1]Schepens Eye Research Institute, Mass Eye and Ear, Harvard Medical School, Boston, MA02114, USA
[2]School of Computer Science and Engineering, Beihang University, Beijing, 100083, China.
[3]School of Optoelectronic Information, University of Electronic Science and Technology of China, Chengdu, Sichuan, 610054, China
[4]College of CCIS, Northeastern University, Boston, MA, 02115, USA,
[5]Institute of Computer Vision and applied Computer Sciences, Leipzig, 04251, Germany

shuhang_wang@meei.harvard.edu, via_alison@buaa.edu.cn, putian@uestc.edu.cn, patwang@ieee.org, pperner@ibai-institut.de

Abstract. Image haze removal has been extensively studied, but there has been no such an image database regarding the haze effect. As a result, it is hard to carry out fair comparison, since different haze removal methods may be useful for different types of images, and meanwhile, it is not convenient for readers to verify the statistical priors that are widely used for haze removal. To solve these problems, we collected a database of more than 4,000 images of different kinds of outdoor scenes. The images of this database are grouped into 4 classes. Along with the database, we also present the analysis and comparison of the amplitude spectra of these 4 classes. We call this the outdoor image database with haze classified. Our purpose is to help develop image haze removal, as well as verify and discovery statistical priors of images.

1. Introduction

Images captured in outdoor scenes often suffer from poor visibility and color shift due to the presence of haze. Koschmieder [1] proposed a hazy image formation model, which has been widely adopted by image haze removal methods [2, 3, 4]. The model can be simply expressed as:

$$I^c(x,y) = J^c(x,y)t(x,y) + A(1 - t(x,y)) \tag{1}$$

where $I(x,y)$ indicates the hazy image captured, $J(x,y)$ indicates the haze-free image, A is the atmospheric air light, $t(x,y)$ represents the portion of the light that is not scattered, and $c \in \{R,G,B\}$ represents the color channel index. The key problem of haze removal is to estimate the parameter $t(x,y)$, which is related to the known scene depth. In other words, it is an ill-posed problem to estimate the parameter

$t(x, y)$ from a single image, and haze removal methods often obtain the parameter based on strong assumptions.

These strong assumptions are usually derived from the statistical priors of outdoor images. To solve the haze removal problem, these methods often tune the properties of hazy images to meet the priors obtained from haze-free images. For example, He *et al.* [5] propose a simple dehazing method using the dark channel prior based on statistics, *i.e.*, most local patches in haze-free outdoor images contain some pixels which have very low intensities in at least one color channel. Nishino *et al.* [6] assumes that the scene albedo and depth are two statistically independent latent layers. To verify these existing priors as well as find new priors, it is required to build a database consisting of haze-free images and hazy images.

In addition, with many image haze removal methods having been proposed, it becomes pivotal and challenging to assess the performance of these methods. In one hand, as we all know, each method is often effective to a special kind of images with similar haze effect. To demonstrate their performance, existing haze removal methods often use a few images as examples. It is unavoidable for a paper only to demonstrate the best results. In another hand, as reported by Ma *et al.* [7], very few works have been done for the evaluation of dehazed image quality, while Ma *et al.* [7] built an evaluation method using 25 hazy images as well as their deahzed versions. We can see that it is highly required to carry out comparison on the same hazy images. Moreover, since it is impossible for a method to be effective for all images, it is reasonable to test the performance on images with different haze effect to highlight the advantage of the method specifically. To our best knowledge, there has not been such an image database.

In this paper, we introduce such a database. This database has two main concerns. 1) The database contains a subset of high quality haze-free images, which can be used to verify and observe the statistical priors of haze-free images. 2) The haze of different images is classified into three classes according to the haze effect, which is convenient for comparing haze removal methods in different conditions. We present in Section 2 the classification of the images by subjective assessment and the statistics of the amplitude spectra. Along with the database, we also present the analysis and comparison of the amplitude spectra of these 4 classes. In Section 3 we analyze and compare the results while we give conclusions in Section 4.

2. Image Classification

Our image database consists of 2000 high quality haze-free images and 1,464 hazy images. Most of the original images are captured by the authors, while some are collected from miscellaneous sources, such as the published papers [4, 5, 8] and the websites [9]. We classify these images into 4 classes regarding the haze by two steps: 1) classifying images into 4 classes by subjective assessment, and 2) analyzing and comparing the 4 classes based on 1/f power law of the amplitude spectra.

2.1. Subjective Assessment

Based on the subjective perception of the image contrast, color saturation, and haze, the image database is divided into four classes: haze free, light haze, mild haze, and dense haze. The characteristics of the images in each class are described as follows. Note that the descriptions are mainly about the non-sky areas:

1. Haze free: images have high contrast, pleasant saturation, and have no noticeable haze (the first row of Figure **1**).
2. Light haze: images have slightly noticeable haze in some areas and the details in most local areas are easily observable (the second row of Figure **1**).
3. Mild haze: images have mildly noticeable haze in most areas and the details in some local areas are observable (the third row of Figure **1**).
4. Dense haze: images have very low contrast and poor color saturation (the fourth row of Figure **1**).

Based on the description of each class, we conducted subjective assessment to classify these images into 4 class. 15 observers have taken part in the subjective assessment. The subjective assessment system is designed as: each image was displayed in a random order, and observers pressed 1, 2, 3, 4 to indicate the class of the displayed image. The classification rule is: an image is classified into the class i if the number of observers choosing the class i is the most. Since different observers may have different preferences, it is possible that an image may have more than one chosen class. In such case, we classify to the image into more than one class.

2.2. Statistics of Amplitude Spectra

In this subsection, we first introduce the 1/f power law of the amplitude spectra of natural scenes, and then show that haze can change the amplitude regarding the frequency, which can be used for image haze removal and rough haze assessment.
Natural scene statistics have been extensively studied and applied by image processing methods [10]–[15]. As one of the most well-known examples, the 1/f power law states that the amplitude spectra of natural scenes is linear with the reciprocal of frequency [16], [17]. Letting $\hat{I}(f)$ be the Fourier transform of an image $I(x)$, the amplitude spectra were found to obey a power law as follows:

$$\varepsilon\{|\hat{I}(f)|\} \propto 1/f \tag{2}$$

where ε is the expectation operator taken over the ensemble of natural images. In the frequency domain this implies that equal power is found in octave wide bands. In the spatial domain this implies that long range slowly decaying correlations exist.

Fig. 1. Examples of the four image classes regarding to haze. The first, second, third, and fourth rows present the haze-free, light-haze, mild-haze, and dense-haze images, respectively.

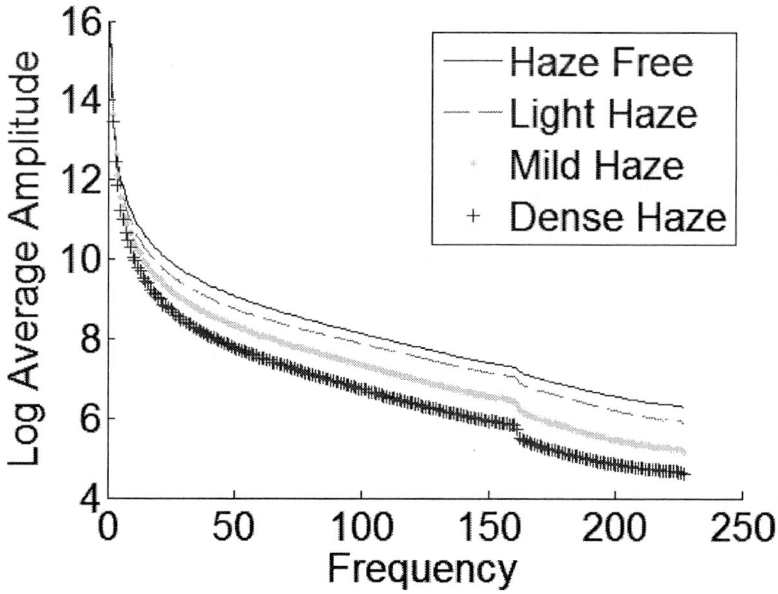

Fig. 2. The average amplitude spectra of the four image classes.

Similarly, we make statistics of the amplitude regarding the frequency for each image class. The average amplitude of the four classes is shown in Figure **2**. We can see that with the increasing of haze, the amplitude value of the four classes decrease. For easy expression, we define the average amplitude in [50, 200] as the average value of each line in the frequency interval [50, 200]. The 95% confidence interval of the average amplitude in [50, 200] are shown in Table **1**. Note that the amplitude value can be used as an import characteristic of haze, but the interval of different classes may be overlapped with each other. There are two main reasons. First, different observers have different preferences, so that they may classify the same image into different classes. Second, different areas of the same image may vary a lot, while the amplitude is calculated based on the whole image. However, the general trend of the amplitude is related to the haze and the frequency.

Table 1. Average amplitude in [50, 200] indicates the average of amplitude between the frequency 50 and 200.

	95% confidence interval of average amplitude in [50, 200]	Image number
Haze Free	[6.75, 8.62]	2720
Light Haze	[6.43, 8.29]	372
Mild haze	[5.79, 7.80]	586
Dense haze	[5.19, 7.70]	555

3. Analysis and Discussion

By examining the testing images presented in the published papers, we found that existing methods often focus on the images with mild haze and simple scene depth variation. It is probably because 1) for images with light haze, it is not necessary to use special haze removal methods to process the image, since traditional image enhancement methods is effective to remove the blur due to light haze; 2) for images with dense haze, existing methods are not often effective, 3) for the scenes with complicated depth change, it is hard for existing method to detect scene depth change exactly, and the areas with different scene depth usually require much different processing. As a result, existing methods are prone to select their best processed images to compare with other methods. In fact, it is not surprising a method is effective to a special kind of images, but it is necessary to figure out the advantage of the method specifically.

For the statistics of the amplitude spectra of the images with different haze, we can see that the amplitude spectra is related to haze amount. Since the light that is diffused or scattered can reduce the high frequencies of images, the larger the haze amounts, the smaller the amplitude value of high frequencies. We take this observation as a prior for image haze removal. Along with the database, we have proposed a haze removal method using the statistical prior of the amplitude spectra, which will be presented in our paper. Similarly, the database[1] can also be used to verify or find more new priors, which can be a guide for developing new methods.

4. Conclusions

Images captured in outdoor scenes often suffer from poor visibility and color shift due to the presence of haze. There is a general hazy image formation model by Koschmieder [1] that is been widely uses. The strong assumptions used for the model are usually derived from the statistical priors of outdoor images. To solve the haze removal problem, these methods often tune the properties of hazy images to meet the priors obtained from haze-free images. To verify these existing priors as well as find new priors, it is required to build a database consisting of haze-free images and hazy images.

Many image haze removal methods having been proposed, it becomes pivotal and challenging to assess the performance of these methods. In one hand, as we all know, each method is often effective to a special kind of images with similar haze effect. To demonstrate their performance, existing haze removal methods often use a few images as examples. It is unavoidable for a paper only to demonstrate the best results.

It is highly required to carry out comparison on the same hazy images. Since it is impossible for a method to be effective for all images, it is reasonable to test the performance on images with different haze effect to highlight the advantage of the method specifically. To our best knowledge, there has not been such an image database.

We introduce such a database. This database has two main concerns. 1) The database contains a subset of high quality haze-free images, which can be used to verify and observe the statistical priors of haze-free images. 2) The haze of different images is classified into three classes according to the haze effect, which is convenient for comparing haze removal methods in different conditions. Along with the database, we also present the analysis and comparison of the amplitude spectra of these 4 classes.

Our study showed that existing methods often focus on the images with mild haze and simple scene depth variation.

For the statistics of the amplitude spectra of the images with different haze, we can see that the amplitude spectra is related to haze amount. Along with the database, we have proposed in this paper a haze removal method using the statistical prior of the amplitude spectra. Similarly, the database[1] can also be used to verify or find more new priors, which can be a guide for developing new methods.

[1] The users can download the image database and the code that is used to calculate the amplitude spectra by visiting this webpage:
https://wordpress.com/posts/shuhangwang.wordpress.com.
In case that the readers cannot download the database successfully, please send email to xiaohangpp@gmail.com to get more information about the database. In addition, we will keep adding more images to the database and updating the statistical results accordingly.

References

1. H. Koschmieder, Theorie der horizontalen sichtweite: kontrast und sichtweite. Keim & Nemnich, 1925.
2. S. Wang, W. Cho, J. Jang, M. A. Abidi, and J. Paik, "Contrast-dependent saturation adjustment for outdoor image enhancement," JOSA A, vol. 34, no. 1, pp. 2532–2542, 2017.
3. Q. Zhu, J. Mai, and L. Shao, "A fast single image haze removal algorithm using color attenuation prior," IEEE Trans. Image Process., vol. 24, no. 11, pp. 3522–3533, 2015.
4. B. Li, S. Wang, J. Zheng, and L. Zheng, "Single image haze removal using content-adaptive dark channel and post enhancement," IET Comput. Vis., vol. 8, no. 2, pp. 131–140, 2013.
5. K. He, J. Sun, and X. Tang, "Single image haze removal using dark channel prior," IEEE Trans. Pattern Anal. Mach. Intell., vol. 33, no. 12, pp. 2341–2353, 2011.
6. K. Nishino, L. Kratz, and S. Lombardi, "Bayesian defogging," Int. J. Comput. Vis., vol. 98, no. 3, pp. 263–278, 2012.
7. K. Ma, W. Liu, and Z. Wang, "Perceptual evaluation of single image dehazing algorithms," in Image Processing (ICIP), 2015 IEEE International Conference on, 2015, pp. 3600–3604.
8. T. Pu, S. Wang, and P. Wang, "A perceptually inspired method for enhancing contrast in uneven lighting images," in Man-Machine Interactions, 2017 International Conference, 2017, p. (accepted, to appear).
9. "Nipic." [Online]. Available: http://www.nipic.com/index.html.
10. K. Moorthy and A. C. Bovik, "Blind image quality assessment: from natural scene statistics to perceptual quality," IEEE Trans. Image Process., vol. 20, no. 12, pp. 3350-64, 2011.
11. M. a Saad, A. C. Bovik, and C. Charrier, "Blind image quality assessment: a natural scene statistics approach in the DCT domain.," IEEE Trans. Image Process., vol. 21, no. 8, pp. 3339-52, 2012.
12. Y. Fang, K. Ma, Z. Wang, W. Lin, Z. Fang, and G. Zhai, "No-reference quality assessment of contrast-distorted images based on natural scene statistics," IEEE Signal Process. Lett., vol. 22, no. 7, pp. 838-842, 2015.
13. L. Tang, L. Li, K. Gu, X. Sun, and J. Zhang, "Blind quality index for camera images with natural scene statistics and patch-based sharpness assessment," J. Vis. Commun. Image Represent., vol. 40, pp. 335-344, 2016.
14. B. Zhang and A. U. Batur, "Illumination estimation using natural scene statistics." Google Patents, 2016.
15. L. Li, Y. Yan, Z. Lu, J. Wu, K. Gu, and S. Wang, "No-Reference Quality Assessment of Deblurred Images Based on Natural Scene Statistics," IEEE Access, vol. 5, pp. 2163-2171, 2017.

16. D. J. Tolhurst, Y. Tadmor, and T. Chao, "Amplitude spectra of natural images," Ophthalmic Physiol. Opt., vol. 12, no. 2, pp. 229-32, 1992.
17. E. P. Simoncelli and B. a Olshausen, "Natural image statistics and neural representation," Annu. Rev. Neurosci., vol. 24, pp. 1193-216, 2001.

Author Index